Praise for

CALLOUSED HEART

"An intentional, raw, and truly authentic dive into questions as old as humanity itself. The story of Calloused Heart, isn't just a story about a Marine and his continued journey through relationship with Christ; it's your story, it's my story, it's our story. From first page until the last you can feel the concussion of the shells, connect intimately with the loss, and ruminate with questions mused by the author's life.

This is not just an incredible story, but an important and timely read. One that could not have come at a more important moment in our journey as a nation. Steve has delivered a unique and epic conversation worthy of all who desire to better themselves, deepen their faith, and process their true identity."

–John Beaty, Police Officer, City of Los Angeles (LAPD)

"This book delivers a profound impact. Steve has captured, in words, the walk of finding his identity in Christ, even after witnessing and being in situations that would leave a person numb and trembling. Transformation through suffering is the model of Christ. He is also a model of joy, community, and restoration. Steve and this book are evidence of all. Well done!"

–Kathleen Arai, Ph.D., Psychology

"Knowing and following the will of God can be a challenge for any person calling themselves a Christian. But, what if God's will for your life calls you to be a warrior. What if the call on your life is to protect your country as a soldier or your community as a police officer? Calloused Heart is a great read for any person beginning the process of self-reflection and assessment required to be both a warrior and follower of Jesus Christ. Both come with a cost."

–**Roger Johnson,** Chief of Police (Ret.)

"My friend, brother in the Lord, and fellow warrior Steve Sanderson has done a fantastic job at fusing his real-life experiences with the truth he lives by to answer many of the questions that warriors and those connected with them struggle with. This book is a beautiful body of work designed to give the reader insight into things they may or may not have experienced, but that humanity has wondered about for hundreds of years.

I have used many of these principles to survive both internally (spiritually) and externally (physically) over the years. Therefore, I can bear witness to things spoken of in this book as Steve shares his experiences. I hope you will read this book with an open mind and a curious heart and that it ultimately serves as a tool to help bring you closer to the peace that surpasses all understanding. The world desperately needs faithful warriors."

–**Byron Rogers,** (USMC), Executive Protection, Author, & Founder of Protector Nation and the League of Executive Protection Specialists

Calloused Heart

CALLOUSED HEART

NAVIGATING THE BALANCE BETWEEN FAITH AND VIOLENCE

STEVE J SANDERSON

ALLEGIANT

ISBN: 979-8-9855851-0-0 (hardcover)
ISBN: 979-8-9855851-1-7 (softcover)
ISBN: 979-8-9855851-2-4 (ebook)

Published in the United States by ALLEGIANT

Book and Cover design by Steve J Sanderson

callousedheartbook@gmail.com
www.callousedheartbook.com

10 9 8 7 6 5 4 3

For my parents, Robert and Marie

Pain and Suffering are always inevitable for a large intelligence and a deep heart. The really great men must, I think, have great sadness on earth.

– Fyodor Dostoevsky

CONTENTS

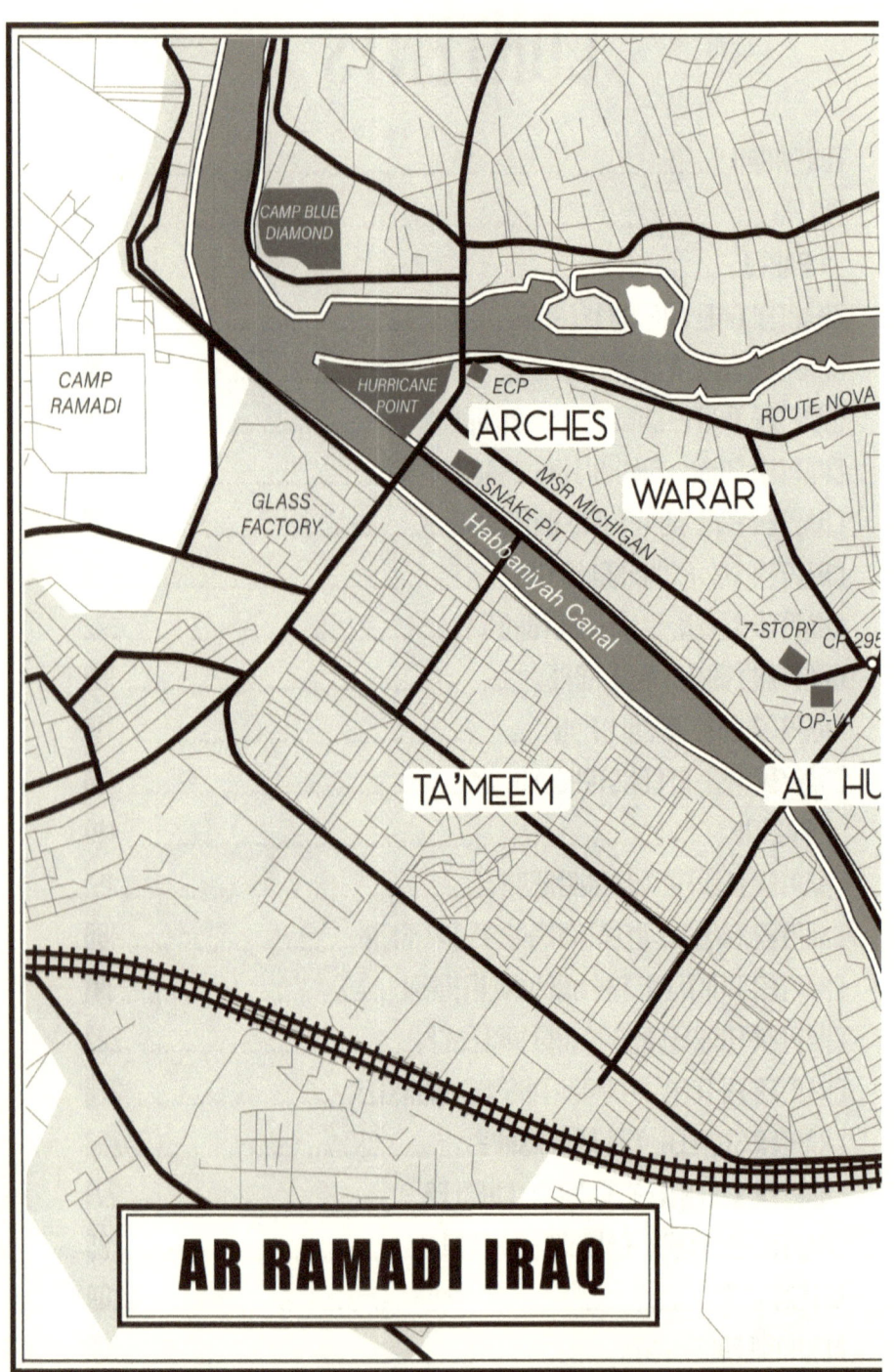

CAMP BLUE
DIAMOND

CAMP
RAMADI

HURRICANE
POINT

ECP

ARCHES

ROUTE NOVA

MSR MICHIGAN

WARAR

GLASS
FACTORY

SNAKE PIT

Habbaniyah Canal

7-STORY

CP 295

OP-VA

TA'MEEM

AL HU

AR RAMADI IRAQ

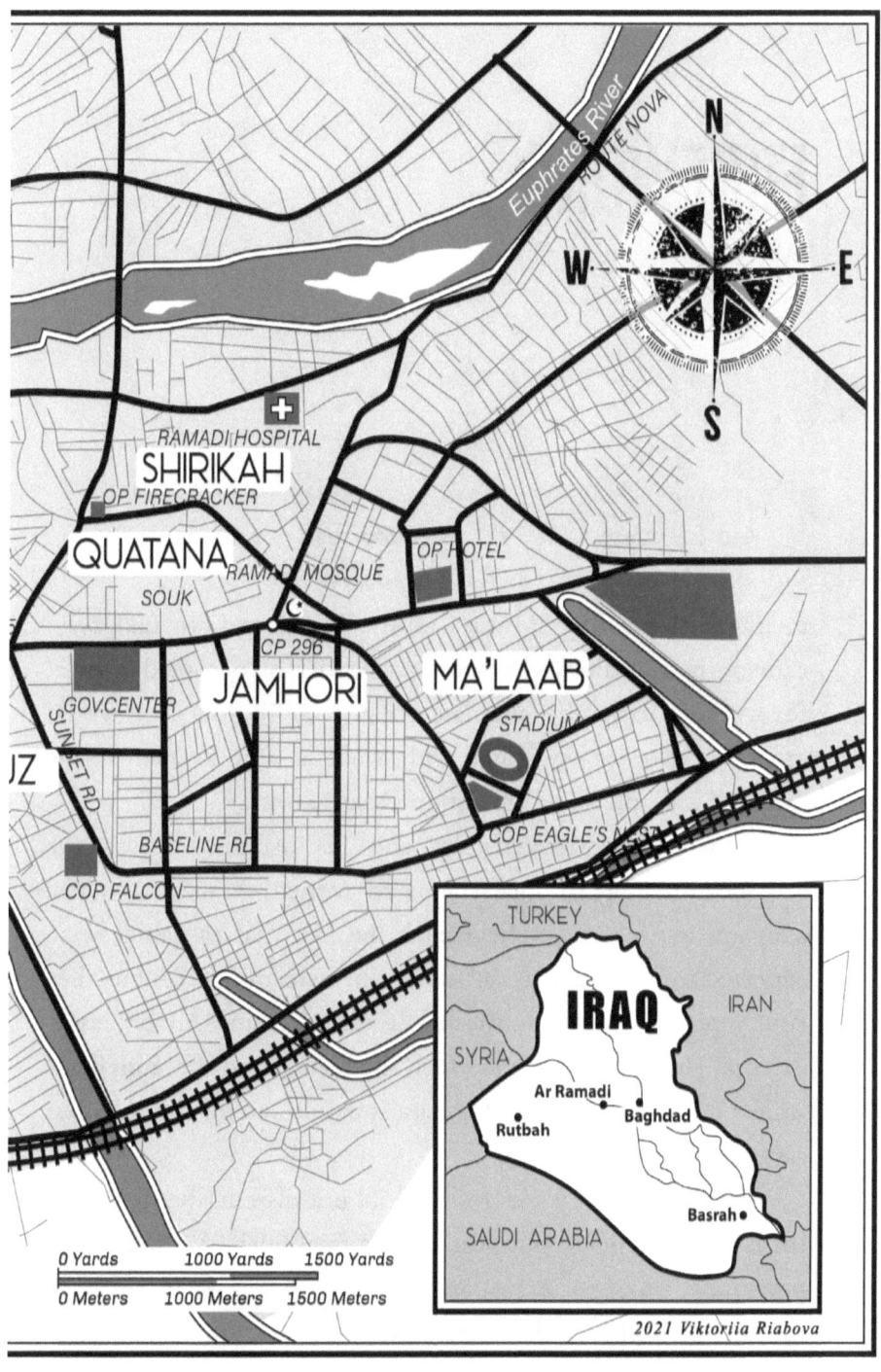

ROUTE NOVA

Euphrates River

N

W   E

S

⊕ RAMADI HOSPITAL

SHIRIKAH

OP FIRECRACKER

QUATANA

RAMADI MOSQUE

OP HOTEL

SOUK

CP 296

GOV.CENTER

JAMHORI

MA'LAAB

STADIUM

ˈUZ

SUNSET RD

COP EAGLE'S NEST

BASELINE RD

COP FALCON

TURKEY

IRAQ

IRAN

SYRIA

Ar Ramadi

Baghdad

Rutbah

Basrah

SAUDI ARABIA

0 Yards   1000 Yards   1500 Yards

0 Meters   1000 Meters   1500 Meters

2021 Viktoriia Riabova

FOREWORD

On February 22nd, 1991, a young man commented, *"My life will never be the same. I've just glimpsed my own death."* He was a member of my Fire Power Control Team, and we had been running patrols along the southern border of occupied Kuwait for weeks. I remember the date because, in 48 hours, we would cross that border en masse with what is now known as the "ground war" phase of Operation Desert Storm.

I had just provided him and his teammates with the classified briefing of the scheme of maneuver and our initial assignments. I had not held back the fact that casualties were expected to be as high as 50%. The entire team fell silent as they contemplated what this meant. It didn't require much imagination. They had personally seen the terrain and the minefields while providing reconnaissance and surveillance for the brief but savage battle at Khafji.

To their credit, it was resignation, not dissent, that fell upon them as we prepared our gear for the final push. A few months previously, the most important thing this young man had been planning was what to do on his weekends, and now he could not be sure he would live to the next weekend. He was maturing far faster than his friends back in the United States while he grappled with his own mortality.

The fallacy of youth is a perception of immortality, but what happens when that perception is shattered? No one so close to death will ever accept life on someone else's terms. They have been forever changed. Questions like "Who am I?" and "Why am I alive?" are no longer abstract concepts but profoundly personal and vitally important. Nor, when one's death seems imminent, can the answers be postponed. No cost or effort is too great in exploring every facet and nuance.

What we've been taught and what we truly believe are a lot farther apart than most people realize. "Does God really exist?" "Is there life after death?" "What is my purpose for being?" I knew exactly what he was experiencing. More than twenty years earlier, I had experienced the same feelings in Vietnam. Nor are they unique or contemporary. Thousands of years earlier, the Greek philosopher Socrates is reported to have stated that "The unexamined life is not worth living." I believe that nearly everyone experiences these feelings at some point in their lives, albeit few so young in years and maturity.

Combat is not something that can be explained. It has to be experienced. Perhaps that is why combat veterans have such a difficult time talking about it. There are no objective metrics. It is deeply personal. It has been said that combat has produced more poets and philosophers than academia ever will. The search for meaning in life is not an intellectual endeavor. It is one of the deepest soul-searching.

Combat strips off the trappings of culture and society, revealing the essence of a person. Social norms and learned behaviors like tact, manners, etiquette, and propriety, are easily demonstrated, whether or not they accurately reflect our true character. Indeed, the only character trait that can't be faked is courage, and in combat, that is immediately conspicuous. At the barest, even the soul is evident.

Life and death are no longer general concepts but absolute and unequivocal. In times like these, God is no longer an abstract generality. God becomes real. For those who have experienced combat, developing a genuine belief in God is not a quantum leap of judgment but a single step of faith. It provides hope in a reality predisposed to despair.

The adage, "There are no atheists in a foxhole," is no longer a cliché. Notwithstanding, a true belief in God is not one that can be accepted at face value. It is an intensely intimate question that begs for rumination and introspection. Irrefutably, we must acknowledge an afterlife or accept the futility of this one. Without these beliefs, our very existence becomes happenstance and pointless.

This book is about another young man, in a later war, but forced to come to grips with the same issues. The first part of the book is written in the first person and is in a journal format. It reads like a novel, and the reader can't help but be drawn into his experiences, especially those that foreshadow the challenges to come. Preparing for your death is certain to invite exploration of your life.

By anyone's definition, writing your own eulogy certainly has to qualify as thought-provoking. Thinking of how you want to be remembered at an age where most have not even considered that they will die is a sobering experience. Likewise, losing your girlfriend just before deploying, creating a will, assigning powers of attorney, and realizing that you have literally nothing of value to bequeath, provides a perspective that few people will ever experience at such a tender age.

If the account stopped here, it would be an exciting story, but just a story. The second part of the book is an intimate examination of the challenges and conflicts that have shaped and continue to influence his life. Using both historical and personal examples, Steve Sanderson confronts the hard questions, such as, "How have these experiences changed me?" "Who am I now?" "Am I defined by my experiences, or can I rise above them?" "What is my purpose for being alive?" "Can a Christian be a warrior?"

As Steve works through his doubts and fears, he becomes aware of the staggering difficulties in becoming the person he wants to be. In seeking answers, he examines his own life, his values, beliefs, and priorities, especially what he has been told compared with what he has seen and experienced. He recognizes that he needs to be loved to be complete, not only by God and his family, but an equal partner and helpmate.

Not surprisingly, he is afflicted with PTSD and must now battle a pernicious and pervasive enemy that is not fully understood and easily underestimated. Feelings of isolation, sadness, guilt, shame, and worthlessness, are met with anger and detachment, damaging old relationships, and making new ones nearly impossible. Confronting these fears is the most difficult of all.

After leaving the Marine Corps, Steve finds himself lost in the world he had left. He's not the same person. He feels detachment and loneliness from those who no longer understand him. It is during this period that he again seeks God as the one who most understands and least condemns. The things he sees as important are not the same as before. The luxuries of the world are not as appealing as they once were. He hears platitudes that he knows have been adopted and recited without reflection.

They are meaningless drivel. He feels disdain for those who indulge their selfish desires at the expense of others. He is judged lacking because he distinguishes evil from violence.

FOREWORD

Violence is not intrinsically evil and is sometimes a necessary tool to protect the innocent and suppress evil. How can he make them understand that a free will incurs a moral obligation to stand against evil? When evil is not confronted it proliferates unchecked.

This is not a "how to" book for readers to find meaning in life. It is a narrative of how Steve found meaning. While such questions are always personal, the answers lead to a relationship with God that provides the source for peace and hope. Hope is the antidote for despair because hope and despair cannot coexist. Ironically, we haven't really lived until we've found something worth dying for.

I have been honored to be asked to write this forward. Like Steve, I struggled finding my place in the world until I accepted that God is far more practical than I had given him credit for. Once I accepted that my calling was in protecting the weaker against those who would victimize them, I found the peace and comfort of following God's will.

I can assure you that finding the answer to the meaning of life is not as important as finding the answer to the meaning of *your* life. None of us can save the world, but we can each make a difference. Not everyone is suited to be a warrior, but the Scriptures are clear that God uses warriors like Joshua, Samson, Gideon, David, and many others, for that very purpose. It is just as clear that each of us is individually called based upon our own talents, strengths, and interests. I pray that you seek until you find what makes *you* important and why God has a plan for *your* life. An answer to prayer is a glimpse of Heaven.

– Charles "Sid" Heal
Author of *Concepts of Nonlethal Force* & *Field Command*
July 20th, 2021

PREFACE

How can an all-knowing, all-loving, all-powerful God allow suffering to happen? Humanity has struggled with this problem for centuries. It is known philosophically as the *Problem of Evil*. Writers and thinkers such as G.K. Chesterton, C.S. Lewis, David Hume, Fyodor Dostoyevsky, St. Augustine, and many others dating back to Epicurus have wrestled with this problem. This problem has drawn many to God to restore some kind of hope for humanity as well as driven others to abandon the possibility of God's existence. This problem raises the question of what Christian participation in violence should be.

The purpose of this book is to share my experiences within hostile and passive settings in order to better understand the relationship between God and violence. The book is written in two parts: Part I is a creative non-fiction narrative dialogue. It covers my time throughout the Marine Corps and deployments to Ramadi, Iraq. PART II covers my post-traumatic growth process after the military written in a conversational tone. There are eight specific stages of spiritual development present in my journey that are covered here. The underlying themes of the book are the three pillars of Christian faith: Identity, Purpose, and Belonging.

The narrative tone of Part I changes throughout the story to follow the progression of events and how they affected me from 18-22 years old. The words and tone parallel the writing of my deployment journals at the time the events happened. One thing I became aware of as I reread these journals was that both of them read differently—as if written by two different authors only a year apart.

It was eye opening to observe the trauma that affected me and provided a reference point for a significant personality change within only a year's time. There are some things about God that became clear in moments of pain but most of what I've come to understand is a result of my post-traumatic growth years. Not everything said in the dialogue of the story are beliefs I still hold, but my intention was to capture a maturing process through the eyes of a young Marine as accurately possible.

I sometimes use Psychology to explain theological points throughout Part II. While it helps us understand human behavior, I do not see it as necessary for spiritual maturity. To do so would violate the authority of Scripture. Psychology has provided a tremendous amount of knowledge on human motivation and behavior, but I am careful to use illustrations of human development not methodologies for fulfillment. Only God's Word can do that. Like any science, psychology changes with periodical revisions to the Diagnostic and Statistical Manual of Mental Disorders (DSM). God's word is unchanging.

Scripture is the ultimate lens in which I view the world and psychology. Psychology is just one venue God has used to illustrate His marvelous works, though without it I would not be deprived of biblical understanding. I am not a Christian because it makes sense, I am a Christian because I can't deny what I've seen God do in my life. God's word is truth even if it doesn't make sense. My goal in using illustrations is to bring clarity as you may move closer to truth.

Additionally, all parallel texts augment the biblical references used, not the other way around. Not everyone will have a childlike faith to believe what Scripture tells us. Many of us are understandably skeptical. The older we get the less we are inclined to take things at face value. There is no shortage of deception in the world and no one wants to suffer for unnecessary mistakes. God gave us an incredible creative capacity so that our exploration would reveal that God's truth always has been and always will be.

The beauty of God's grace is that He is always able to appeal to our understanding. This is why science and psychology can be conduits which lead us to God's truth even though they are not essential. Sometimes people need to experience some-thing before they can believe it. John 20:27-29 says, "Then Jesus said to Thomas, 'Put your finger here and observe My hands. Reach out your hand and put it into My side. Don't be an unbeliever, but a believer.' Thomas replied, 'My Lord and my God!' Jesus said to him, 'Because you have seen Me, you have believed; blessed are those who have not seen and yet have believed.'"

When it comes to the problem of evil, I believe that violence is a tool for justice in the right context. I do not believe the issue of suffering necessitates a distant or non-existent God. A gal I met shortly after the military asked me, "How can you be in the military and call yourself a Christian?" She believed the two were contradictory positions. I can't blame her for having an aversion to violence but I do hold her accountable for calling herself a Christian and letting politics and emotion speak over someone's decision to serve. Instead of trying to understand how violence can be a tool for justice, she gave into fear and ran in the opposite direction.

It's human to desire peace and want to avoid conflict. Violence and suffering take things from us that will never be replaced. They distort our reality and tear us from the foundation of our design. Suffering causes us to seek refuge outside of the world we cannot escape. I hope these pages will help you navigate the balance between faith and violence with a better understanding of the judicious use of force. It is with a hopeful heart that you will find meaning in suffering and continue to face the world with an endless capacity to love.

God is not separate from violence and pain. He is the God who understands and participated in our suffering (Romans 8:26). God works in unconventional ways, pursuing us relentlessly with limitless patience to offer joy in all circumstances. His love is unconditional and stands above our circumstances so we can know Him and find joy wherever we are. In every trial of life, know that God is good all the time.

Santa Barbara, California
August 2021

PART I

CHAPTER ONE
THE AWAKENING

IT WAS MY FIRST day as a sophomore in high school. I ungraciously rolled to the floor preparing for another academic year. Friends across town were waking up to their senior year with a signed contract for the Marine Corps. I envied their departure from our small town. I muscled through my usual routine of urinating half asleep before showering. I paced for the kitchen through my parent's room where my mother stood suspiciously fixed to the television. I glanced to see what held her attention this early, finding myself curiously engaged.

Flames consumed the North Tower of the World Trade Center, as viewers tuned in with mutual confusion. Reporters rambled to fill the empty space, composing a soundtrack of concerned assumptions. Smoke continued to bleed from the tower, adulterating the air with unrestrained sorrow.

"What started the fire?" I asked. "Shh, I don't know." She responded, without breaking from the screen.

"Did they say what happened?!" I prodded.

"I don't know, I just turned it on." she closed.

She walked out to continue her morning ritual. I snapped back to investigate just as the second plane hit the South tower. Confusion escalated to fear. Reverence began forming in terror, silencing my juvenile opinions. *This cannot be an accident.* At fifteen I knew something was wrong. An ember ignited a flame in me as dark grey billows of smoke veiled the upper levels of both towers. I was lost in what this meant–what this would do to the United States.

I made my way to class with the news coverage as an afterthought, but every classroom sat affixed to the unfolding story on television. Two more planes diverted their course, one crashing through the western wall of the Pentagon. The other into a Pennsylvania field. Terrorism was suggested when a second plane collided into the southern Trade tower, changing everything. The nation was suffering the biggest loss of life on its own soil since Pearl Harbor. We would be looking for blood.

Uncertainty loomed like the black clouds ravaging the towers. Anger surged through my veins fueling the desire for vengeance. The thought of joining the Marine Corps grew even more appealing. I found meaning in my future service and I would go down in the pages of history. It was my chance to serve with a purpose and stand taller with a sense of accomplishment. My heart was consumed with fire and ventured into the darkness of that New York sky. My time was coming.

ENLISTMENT

"You're doing what?" Kristi screamed at me when I told her I was joining the Marine Corps. "I can't believe you would do this to me!" She retreated before I had the chance to explain myself. I forced my way into the room to disrupt her protest,

"Do this to you? Why is this decision dependent on you?"

"I thought you cared about us?" She associated me.

"So now it's about us?" I defended.

"I told you I didn't want you to join the military. I'm never going to see you, and when I do you won't be the same person. Why would you leave me like this?" she projected her own intentions. She withdrew her affection by manufacturing indignation of which I was responsible.

Immaturity suspended her judgment until her pressing need for gratification considered the time we still had before I shipped off. She conceded though her outrage was off color. She had been aware of my plans a few months into our relationship. I cared deeply for her and reflected on my decision. It left me wondering if our relationship was worth the fight.

The distance would be difficult, though her insecurities were plotting. Our relationship would eventually come under trial with or without military service. My heart kindled under a new flame, as hers did the same. Good men would be called upon and my loyalty was being tested. I was ready to become something more–to pursue my faith into the unknown. This was a choice between her and God. The pursuit of my identity in Christ might tremble through sharp iron but her banter was an empty breath against the match. This was my journey and I was learning how to start a fire.

Instead of waiting for their phone call, I arrived at the recruiter's office asking for Mephistopheles himself. Academics would give way to the real world. Marine recruiters capitalized on the bloodlust of graduating seniors. September 11th filled quotas with ease. The conflict overseas extended from Afghanistan into Iraq, requiring the need for additional Battalions of Marines. My naïve heart was full of grand intentions yet unprepared for true sacrifice.

I signed the contract for Marine Corps infantry five months after the United States commenced Operation Iraqi Freedom (OIF). With the legal consent of my parents, I was able to sign my name at seventeen years old. They co-authorized their guilt if anything were to happen. High school graduation came faster than I could appreciate and moved me closer to the mark. A high school diploma was the final condition of enlistment.

Families took their seats under the open sky and suspended sun, honoring the graduating seniors in their unsuspecting transition of life. I stood alongside my fellow graduating seniors and Mike Turrell, who would be shipping out to basic training with me. From grade school through high school we were inseparable–just skateboarding across town before we had driver's licenses and burning CDs with downloaded music. We fed off the stories returning back to us from Iraq through friends who deployed ahead of us. We were eager to fight.

Our small town faded away like our patience. Rich green colored gowns and gold tassels draped over each row of graduates, waiting to be called up in our proudest moment of academic achievement. One by one names were announced by the setting summer sun and generous weather. Parents watched their children march with pride into an unknown life of responsibility. For Mike and I, it would be a swift transition from diploma to orders.

Two months after graduation I was standing on the infamous yellow footprints at Marine Corps Recruit Depot (MCRD) in San Diego for basic training. I traded my cap and gown for a set of digital MARPAT (Marine Pattern) fatigues. Thirteen weeks of rigorous training and strict obedience to orders felt longer than my senior year of high school. The physical, mental, and spiritual testing of my character perspired in the lands and grooves of the unique fingerprint of my developing identity.

Drill Instructors (DIs) lined up in front of their respective platoons in Golf Company, reciting their creed. I sat arrested with uncertainty as did the rest of Platoon 2021. Tension gripped the room, silencing passive voices. Finally, chaos ensued when the DIs broke formation screaming in every direction. Their authority would reign for the next three months. Academics were finished.

Recruits from various backgrounds from all over the country were stripped of their personal identities and freedoms. The fight for sanity turned one recruit against another. Every move we made was wrong in the eyes of a DI. One recruit's mistake was everyone's pain. The entire platoon was punished for an individual's crime. Divisions among recruits made punishments more severe and eventually led to compromise. Differences were set aside to create group cohesion, minimizing punishment. If we were going to suffer, we would suffer together.

I pulled my weight to stay off the radar, but my thin frame and oversized uniform stood out with a service record of "03-Open Contract." I was the last candidate expected to fit the image of an Infantry Marine. Grunts embodied the image of the Marine Corps in the public eye but I was far from the stereotype. All recruits with a contract beginning with '03' were drilled harder and longer than the rest of the platoon. The Infantry was not a place for the mentally weak and the DIs knew it.

Basic training broke down into 3 phases. Each phase was structured with a psychological process, beginning with the initial breakdown of self, build up in 2nd phase, and responsibility in the final phase. The initial breakdown removed individuality through uniformity and taught basic principles of rank structure, history, and general orders. The buildup phase moved outdoors to field training and range qualification. This was the first taste of the infantry playground for the 03 types. A healthy dose of sleep deprivation, a rationed food supply, and the notorious "Crucible" hike where recruits traditionally became Marines was our final field exercise. The final phase tightened up our drill movement, uniform regulations, and unit cohesiveness.

Golf Company prepared for graduation after what felt like a year of training. Drill movements on the parade deck reflected our discipline and echoed with generations of footsteps long before our fathers. Drill Instructors pulled cadence from strained voices, never ceasing to highlight faults in the fatigue of our movement. We moved with purpose, every step closer to the title we longed for. We all made a commitment for different reasons and labored through friction in order to become something greater than ourselves. We sacrificed our individual identities to be counted among the few.

Our pride rose like the sun on graduation day, blinding us to the fear and hostility that reigned for 3 months. Family members packed bleachers again, this time to experience the transformation from civilian to Marine. The Drill Instructors who tore us down, strengthened, and led us through recruit training relinquished their command as they placed the Eagle, Globe, and Anchor in each of our hands. With tears in our eyes, they addressed each man in formation with a final "congratulations, Marine."

SCHOOL OF INFANTRY

I received orders for the School of Infantry (SOI) shortly after basic training. All infantry Marines were sent to SOI for further field training and weapons knowledge. Non-infantry

Military Occupational Specialties (MOS) were sent through a condensed version of our field training as their specialties prepared them for indirect combat roles. The core of our training consisted of troop movement, weapons handling, land navigation, and warfare tactics. Instructors were Sergeants and Staff Sergeants who previously served their initial enlistments as grunts. They were hardened leaders.

I grew in maturity as a Marine, though my oversized cammis and juvenile posture carried over into Infantry training. Nothing about a 147lb, six-foot two-inch Marine telegraphed a threat aside from my digital MARPAT uniform. At eighteen, I looked barely old enough to drive and got away without shaving for days at a time. Private First Class (PFC) Chad Oligschlaeger capitalized on potential insecurities, exploiting them in true extrovert fashion. "Dang Sanderson, you're too pretty to be a Marine. I'm gonna call you princess from now on," he declared while in line for the chow hall. Everyone called him "OG" either for the sake of brevity or illiteracy, but likely a bit of both.

Others caught on and joined in during communal roastings. I withdrew my defense at the threat of further pressure. His comment gained momentum and the only way to deflect it was to accept or enhance it. Defensiveness in a crowd of Alpha males signaled insecurity. It destroyed the bond of male trust as it was simply a test of character. I was anxious to evade the spotlight, yet saw the culture for what it was. Ridicule was simply a form of initiation. It was easier to be accepted by showing resilience than not being able to take a joke.

There were no personal feelings associated with insults—we just wanted the freedom to speak our minds. I embraced the culture at the expense of my image, strengthening camaraderie with other Marines in the process. Lance Corporals (LCPLs) Cory Mince from California and Josh Mattfeld from Wisconsin became close friends during SOI. I offered Josh a home on our weekends off and took up surfing with Cory and his brother. Josh appreciated our California adventures and beautiful coastline. Surfing was our way to decompress and maintain sanity outside of uniform. Relationships alleviated the inconveniences of military life. More importantly, they provided a distraction from the fear of the unknown awaiting us in country.

THE FLEET

Cory, Josh, OG, Mike and I all became Mortarmen and graduated from SOI receiving orders for 3rd Battalion 7th Marines (3/7) in 29 Palms. Upon arrival, we were assigned to Weapons Company. Further challenges welcomed our arrival in the high desert, revealing why Marines hated the place. The isolation of the high desert left most of us without vehicles. Barbershops and bars were the only local fixtures accessible. Naturally, liquid diets became the elixir of a Marine's new desert life.

It was rumored that 29 Palms was once an Army installation, but declared uninhabitable before the Marine Corps acquired it. OG thrived on rumors and considered it a testament to the hardiness of being a Marine. Lake Bandini was further insult to injury—the name being an esteemed satire of a lowly sewage plant. The rotten smell lingered across the base, even more so through the plumbing in the barracks. Even the air was a reminder that everything was oppressed for a Marine in 29 Palms.

Our arrival elevated the status of existing Marines in Weapons Company to "Seniors." We were considered "Boots" as junior Marines. The swift rejection from seniors made us outsiders. We were unknown and wouldn't belong until we earned the right. A tradition of hate and discontent through generations of combat vented itself to establish dominance. It wasn't personal just our turn.

Senior Marines assessed our peer group for the roles we would fill. They would tell us who we were and take responsibility for our success or failure. Corporal Shane Burge, a Mortarman from Kansas and senior Marine, selected me to become a Forward Observer (FO) in charge of 81mm mortar Call for Fire (CFF). I wasn't sure if the responsibility was a blessing or a curse, but it was a chance to prove myself and feel less foreign. Miraculously, Mike was chosen to be the second FO from our peer group.

I held the title of Marine but still lacked substance. The separation from home was deteriorating my heart. My attitude became abrasive. I thought about Kristi and the weekends I'd be able to spend with her before deploying. Home was less than two hours away and a blessing few were afforded. I had more than enough incentive to face the challenges ahead and considered those less fortunate. Surfing with Cory and Josh became a weekend priority when field training wasn't scheduled. It was the only chance to have an individual identity.

Mike and I carpooled home to Los Angeles on Friday nights to get away from senior Marines and working parties. We brought other Marines home like liberated prisoners. One risked becoming a casualty to a drunk Marine or becoming the designated driver for a Senior Marine if spotted over the weekend. A handful of us junior Marines didn't drink and became the unofficial DD's. Like Cory, Lance Corporal Matt Beard was also a devout Mormon who never touched alcohol. I was a straight-edge Christian making the three of us the go-to drivers. Upright character became someone else's utility in the fleet.

Only a handful of junior Marines were legally allowed to drink and the rest did anyway. Attempts to control underage drinking were undermined with contempt. Alcohol was the least of our worries with a hostile deployment at hand. PFCs Michael Penney from Michigan and Andrew Bedard from Montana taunted our sobriety by leaving empty beer cans in our rooms to clean up. Marines drank each beer as if it were their last knowing it may very well be.

3rd Battalion 7th Marines was scheduled to deploy late summer, giving us 6 months to integrate from conventional warfare sections into Combined Anti Armor Teams (CAAT). We would be conducting mobilized patrols in country with Humvees carrying heavy weapons and anti-armor capabilities. Each platoon had two sections, broken down alphabetically and by a color. I was assigned to CAAT Red Alpha Section. Each section ran four trucks with my place in the 4th vehicle: Callsign "Red Alpha 4."

Corporal Dale Clifton from Louisiana was the Vehicle Commander (VC) and a school trained machine-gunner. He was a proud Southerner who asserted his roots every time air left his mouth filtered through a ring of Copenhagen long cut. Lance Corporal Seth "Willy" Williams from Washington was Alpha 4's gunner. Willy was spinning up for his third deployment and built a reputation as one of the best M2 .50 caliber gunners in the Company. He could hit targets a few hundred yards out from a moving Humvee.

Doc Gonzales was Alpha section's Corpsman, an Argentinian native on his first deployment who was also placed in Alpha 4. PFC Lira was the driver and arrived to the fleet shortly before me. I filled the last seat as a dismount, but was called up behind the wheel when Clifton saw the chance for a smoother ride. He rotated Lira and I to see who had a better hand at the wheel.

"Lira, how do you even have a driver's license? Sanderson, you're up. Let's see if you can do any better than Lira here. You ever driven off-road before?" Clifton said in frustration.

"Not like this, Corporal." I replied while climbing into the driver seat.

"Go easy on the brakes. It ain't that hard." He scoffed at Lira.

It was a competition for the position. I saw it as another chance to build my reputation and earn a place to belong, not realizing how many additional assignments I had unknowingly volunteering for. Lira's erratic driving was a conscious decision rather than a result of incompetence. I missed the hustle and became the new driver of Alpha 4 with a Humvee license. Our individual positions in the truck teams were solidified.

Field exercises consisted of vehicle operations and machine gun ranges. Regardless of our designated MOS's, each Marine was cross-trained to operate the M2 .50 caliber, MK-19 40mm, M240B 7.62mm, and M249 Squad Automatic Weapon (SAW) 5.56mm automatic weapons. Javelins, AT-4s, TOW missiles, and LAW rockets were rarely fired in CAAT team training due to the expense of each system and rare use in the urban environment of Iraq. Somehow, I was selected to fire one of the few Tube-launched, Optically-tracked, Wire-guided (TOW) missiles that Weapons Company was afforded. Lance Corporal Shane Swanberg, a senior Marine from Washington, knew the system best and guided me through its operations.

"Alright Sanderson, just follow my instructions and press here when I tell you. When you fire, you'll hear an electronic wind up for a half second then *POW!* The TOW missile is wire guided and unravels in flight. It's gonna knock you off target, but wait for the smoke to clear and you'll see a red glow heading downrange. Just keep the crosshairs on target and that glow will center itself on your reticle. It's like a video game."

"Roger, Lance Corporal." I said enthusiastically and proceeded to blow up an armored vehicle staged in the open desert. Firing the TOW missile was a highlight of my training, but only served as a temporary removal from the hostility of the senior Marines waiting to see my reaction. Few expressed their envy, but most remained quiet. Clifton couldn't withhold his jealousy, "I don't know why they picked you, Sanderson. You're still a Boot. They should have picked someone more *qualified* if you ask me." He shadowed my excitement. I concealed a smirk to avoid being pulled for a retaliation working party.

Our first few months in the fleet revealed the nuances of Infantry culture. Drinking was a competitive sport, Boots didn't belong, moral character was challenged, and hostility was a training tool. The external world was suspended for two hours in any direction. Sanity had to be found in what we had. Training in the CAAT teams forged stronger bonds with peers and affirmed our decision to become Marines. Signing a military contract was a statement of our personal allegiance. Autonomy was sacrificed for a common objective. Learning to be comfortably un-comfortable was key for our survival. Isolation tested my mind as I fought to remain optimistic.

A reminder of the discomfort was inevitable every time I left home for a Monday morning formation. I missed home as much as everyone else even with the benefit of being able to visit most weekends. We all struggled to adapt through the evolution of training but grew stronger in our resolve. Despite my growing maturity, I failed to be present. Kristi came to mind and satiated the brooding overthought about why we were training. She was a beacon of hope as well as a point of contention. I felt less pressure to conform to all the cynical perversions of the new environment with someone who knew the person I was before. As much as I desired change and felt called into the unknown, I feared losing something.

CHAPTER TWO
A SEARED CONSCIENCE

IT WAS MID-JUNE with the bulk of our field training winding down in preparation for our departure. Only a handful of field operations remained before the unit met the qualifications to deploy. Swanberg caught me on a weekend I couldn't make it home. He often sat outside his barracks room in a lawn chair drinking.

"Come here Sanderson." He called.

Oh great, I thought to myself. "Roger Lance Corporal," I responded and shuffled over to his view. It was the fear every junior Marine had when spotted on the weekend by a senior Marine who had been drinking.

"What time do you have?" He asked, catching me off guard.

"Uh, it's almost 1730hrs Lance Corporal." I said. I stood at parade rest waiting for an inconvenient task such as a beer run or ride to the bar.

He responded casually, "So, it's 5:30pm, cool."

His informality left me confused as did his lazy request. I underestimated his lack of concern for Marine Corps discipline.

He noticed my blank stare, "It's Sunday dude. And drop the Lance Corporal trash too." he continued after an awkward pause. "Do you drink Sanderson?"

"No, I don't."

"Are you old enough to drink? Probably not…"

"No, but I wouldn't drink even if I was."

He took advantage, "Ah, well I'll remember that next time I need a ride."

I was about to walk away but he insisted, "Hey, you live near LA, don't you?"

"Yes, I usually go home every weekend but had to stay this week." I replied.

"Sweet, there's a few girls that want me to come down there but I don't have a car. Maybe I can get a ride the next time you go?" He interjected with a subtle hint of discretion for me to respond with.

"Sure, I don't think it would be a problem." I said. "I thought everyone eventually got a car in 29 Palms?"

"Nope, I don't have a cell phone either. Makes it hard to be located for working parties and other B.S." I laughed as he stepped inside for another beer.

I walked off feeling less like a foreigner. He challenged the senior Marine persona and never lost his sense of the real world. He would be heading out for his second deployment and came to Weapons Company from 1st Tank Battalion. His understanding of how to "play the game" was inspiring. A Marine who could get out of additional duties without detection was considered a "Skater." Swanberg was a catalyst for small victories and had the sharpest skates. His presence in the platoon brought relief to both senior and junior Marines who fell

17

victim to complete institutionalization.

Pre-Deployment leave was at the forefront of everyone's mind. Mild physical training (PT) every morning kept us active and helped us avoid last minute injuries. Administrative requirements soured our remaining time in the States. We toiled through tax documents, finances, legal paperwork, and so-called "yearbook" photos and "personal biographies." I optimistically assumed we would each receive a unit catalog to commemorate our service similar to high school. Sergeant (SGT) Cahalan, Bravo's section leader, was tasked with collecting CAAT Red's paperwork. He perused my biography and handed it back cynically.

"Good, now rewrite it in the third person." He said.

"Sergeant, why are we writing a personal biography in the third person?" I asked for clarification, placing confidence in my twelfth-grade English skills.

"Ha!" He laughed my innocent confusion aside. "Because that's what they're going to read if you die." He was unmoved in his sarcasm.

"So, this isn't a biography, it's a eulogy. We're writing our own eulogies!" I announced in a suspended experience of my own being.

"No kidding!" He shrugged and proceeded to collect the others with a calloused demeanor.

Anticipating our deployment to Iraq peeled away my bearing. My mind was becoming a prison. The ambivalent conflict of choice my parents conferred onto me took refuge in my mind. My life felt like chaff in the wind and the recently seared USMC tattoo on my arm substantiated the conviction. I loitered outside myself asking if I was sure this is what I wanted to do—as if a choice still remained. Fighting to discover who I

18

was and where I belonged was futile next to the threat of death. I was earning my title while at the same time defending against callous deterioration. The path to would involve blood on my hands.

Writing my own eulogy compromised my conscience. I sank heavily with fault. Fear created doubt about my decision. I had to take ownership of the place I was in but I was gaining momentum faster than I could maintain. I became a passenger to my own vessel at the discourse of the Marines beside me. Yet, I felt responsible for the end result soon to be confirmed in writing, from a third person's perspective. My head and heart were at war within me.

Sergeant Carson–Alpha's section leader, formally pulled Alpha together to pass word before releasing us for the night. The smell of Lake Bandini complimented our interest in platoon formation. He made sure our head count was up and recapped on the week's tasks. The mundane banter of his voice reiterating things we already knew gave me a few minutes to mentally check out. My hope was slipping and I despised the place I was in. Senior Marines made life miserable even if I subtly held pride in who they were transforming me into. The friendships formed through trials restored a sense of gratefulness but I was conflicted. Carson's voice pulled me back to reality.

"Alright, last thing for the night…"

Almost done, I told myself.

"Battalion needs to relocate a Marine from Weapons to Lima Company. CAAT Red was the platoon chosen to give up a Marine." He continued, "I know we've all been training together for a while so I'm just as pissed off about it. But I can't do anything to change it. So, this is how it's going to go: If someone wants to volunteer, speak up now. Otherwise I'm going to choose..."

No one dared to move or break gaze from him. The slightest movement would be interpreted as consent. Every senior Marine was secure from selection. The junior Marines knew this and wouldn't dare risk the camaraderie that took months to earn. We suffered together as brothers; we sweated and complained next to each other. Moving to another Company before deployment disrupted unity and isolated that individual to an unfamiliar team. It was the last place any of us wanted to be under the anxiety of our first combat experience.

My heart kicked through my temples like a drum louder with each second of silence. The threat of being distanced from all the personal connections I made was an emotional disruption I was unprepared to face. I worked hard for the place I was in. I was earning my place among the men I respected. *This can't be happening to me*, I selfishly thought in a familiar voice. I summoned an internal stoicism and mentally prepared for excommunication. No one volunteered for anything in the Marine Corps anyway. I remained silent.

"Alright, since no one wants to speak up, I'll choose." Carson ignored the weight of his orders. "PFC Bedard, pack your gear. You're going to Lima." We immediately shifted our eyes in his direction while concealing relief. I felt a disassociated responsibility in the matter as if to plead with him in genuine protest. "Roger Sergeant," Bedard spoke dejectedly. Carson released the platoon for the night, leaving us to resolve the social tension on our own time.

I grew impatient for pre-deployment leave. The stress of our build-up compounded. The last-minute change left me apprehensive. I needed to be around family. I needed to explain things to Kristi. Similarly, junior Marines who were married were anxious to see their wives and children. Lira and his wife would be having a daughter soon and shared in anxious anticipation. I was beginning to understand why Marines drank, but I couldn't bring myself to submit. It was the only control I had over my life and I wore it like a badge of honor. The two-week break would give me time to regain consciousness.

"Alright gents, don't drink and drive, wrap it before you tap it, and don't be UA [Unauthorized Absence] when we get back..." the Commanding Officer (CO) began the Company's safety brief. The same basic guidelines were reiterated by every subsequent rank in a circular windstorm of oxygen bandits. The chain of command's banter rambled on longer than the drive home. I finally made it to LA seemingly a day after being released from our endless pre-deployment safety brief.

I spent my final days with friends and family and let Kristi know I would be around for a few weeks. If I wasn't saying goodbye to loved ones I was with her. My anticipated relief waned in sync with the dying sunlight knowing her fears had manifest. Her somber mood and partial effort to communicate revealed the place of her heart. I failed to step into her pain until my heart was on the line. Physical intimacy replaced conversation, dragging out what couldn't be said. We slept in a burning room as if life would be extinguished by morning.

I was in a place of incredible uncertainty with God, left soaked in shame having compromised my faith while also attempting to enjoy a life I might never have experienced. She was becoming an intangible reality just as she had warned. Still oblivious to the one-sided commitment, I was lost in the space between the present and an uncertain future. I was at war with God before the war of my life. The Author of life was taking back what was his, as I fought for her, and she fought for herself.

I surfed to avoid the convictions of my actions. I didn't know what I wanted anymore. Everything felt outside of my control and left me fighting for authority through sin. I felt enslaved to the contract signed and understood why it was required. I had no liberty to quit. The thought of 29 Palms crushed my spirit. I prayed for more time—maybe something would change. Only a few hours of leave remained. Pre-deployment leave was over as soon as it mentally began. I dragged out every remaining minute like a fleeting commodity, watching the clock with contempt as I gathered my things to make formation in the morning.

At 0600hrs, Alpha section came together for our first formation back from leave. Sergeant Carson called roll to confirm everyone's return. The temptation to return a day late would result in disciplinary measures, but could be reasonably justified with the right excuse. I thought of how badly I wanted one more night home. He began reading names:

"Burge," Carson began.

"Here" Corporal Burge responded.

"Muniz," he continued. "Muniz!" he repeated himself after no reply.

"Yeah." Corporal Muniz said hungover.

"Look, I know we just got back but y'all need to suck it up and get your head back in the game. I expect you senior Marines to set the standard for these younger guys. How about the appropriate response? We're going to be here longer if I have to repeat myself." Carson was becoming frustrated, "I'll leave you standing here all night if that's what you want." No one made a sound.

"Alright then. Sanderson," he continued.

"Here Sergeant." I promptly responded to avoid conflict.

"Lira," Carson called out with no response. "I guess we feel like standing around all day, don't we?" His frustration peaked. I was distracted by his choice of words. He interjected "we" as if he would be involved in the repercussions. He resumed his dialogue, "Does anyone know where PFC Lira is?" We glanced around hoping someone had an answer.

"Someone better find Lira right now," he stormed off.

"He better not be UA." He said under his breath.

One Marine rushed to his room while another called his phone but produced nothing. The lack of response forced Carson to finish roll and mark him as an UA.

We continued with gear inspections and cleaning weapons at the armory. It was the least exciting thing about infantry life. Gear inspections were tedious but needed. Cleaning weapons was the default chore to fill empty time. After pre-deployment leave, it was going to take a few days to recover our attention to detail.

The first week passed and Lira still hadn't returned to the unit from leave. The longer he remained UA, the worse the charges would become. The impending deployment changed everything in his case. If he didn't return soon, he would be deemed a deserter. The demand for unit readiness forced Carson to replace Lira's position in Alpha 4. Lance Corporal John Reis took Lira's position as a Dismount in my vehicle. If it had been any sooner, Bedard might have remained with Weapons Company.

The CO ordered a Company formation the following week to pass last minute updates and inspire morale. As we came together, he ordered us "At-ease" and instructed the Marines to gather in a mass huddle. Each one of us knew we were heading into a fight that would leave profound wounds. As the CO

23

looked around, the tension grew solemn. His posture commanded respect without a word being spoken before he began. He took a breath as our attention focused. He opened with sincerity,

"Gentlemen, I want you to look to the person on your left, and on your right…"

The entire Company of Marines was taken back momentarily. We glanced around confounded with his request before a feeling of consternation seized us.

He resumed, "One of you is not coming home."

The sentimentality in the silence following his words pierced our hearts where months of training diligently hardened. His words solidified reality. Senior Marines remained stoic in the face of adversity. The nature of our deployment unraveled throughout his speech:

"We are in the business of killing, and business is good. Battalion has confirmed our fight in Ramadi, Iraq. We will be relieving 1st Battalion 5th Marines at Hurricane Point. The fierceness of Marines has driven the insurgency from Fallujah to Ramadi. Ramadi stands as the capital of the Al Anbar province and current Al-Qaeda stronghold. With the aid of foreign fighters and various terrorist groups, Ramadi has become the most dangerous city in Iraq at this time.

We will conduct daily patrols in our AO [Area of Operation] and serve as the Battalion's QRF [Quick Reaction Force]. Our mission is to win the hearts and minds of the Iraqi people. For some of you, this will be your first deployment. For those who have already been downrange, I expect you to lead these Marines from the front. Know that as Marines of 3/7, there is no better friend, and no worse enemy."

His words were a reminder of why I joined the Marine Corps. I didn't choose the infantry for financial security or a routine schedule. I chose to give up convenience for the chance to become something. The fight ahead carried the blood that would christen my journey. I was empowered by the knowledge that I was not alone. The fear of dying lost power in that moment. Each one of the Marines next to me was scared, though each one was still human. We faced our fears for the love of our country, to continue a family tradition, to find our tribe.

It wasn't money or a chance to travel the world that substantiated our call to service. We willingly chose the path of sacrifice in exchange for meaning. We joined the infantry to fight, to sweat, and bleed over what we believed. We chose to sacrifice our body and mind to fulfill a purpose. The lives next to us depended on every individual action. We stood for each other at the cost of our most valuable gift.

I remained faithful by choice, not obligation. I was emboldened, though fears still lingered outside the motivational briefs. Stories of honor and courage vindicated my decision to join through the remainder of his speech. Many men died in foreign wars but faithfully chose to answer the call. This was my time to serve the way the Marines on the shores of the Pacific Islands, the snow-covered valleys of Korea, and the jungles of Vietnam had done. I would leave my mark on the sands of a Middle Eastern desert. I owned my war and was defining myself as a Marine.

I felt powerful in my new persona as the social climate consumed me. I was an apprentice in the business of killing. My time for vengeance was near. I grew to hate an enemy I did not know and remembered the World Trade Center. Courage was attractive but I still lacked wisdom. Emotions conflicted with reason when motivation faded. Doubt raged at the selfish desire to stay. I underestimated the sacrifice of my decision. I could lose everything in my pursuit. I was a ghost cloaked in an impeccable uniform.

I thought of Kristi at the call to embrace my violent potential. The thought of her flashed as a warning of who I was becoming. The fear I caged broke free vindicating her protest. I saw the person I used to be, in the same suspended experience while writing my own eulogy. Our relationship drifted and confirmed her concern. I raced home on a final weekend pass hoping to satiate the void consuming my heart. She seemed right even if she was selfish.

I barreled out the front gate of 29 Palms headed for LA. I set out to restore balance in our relationship before leaving. I saw the reflection of the person she feared and shared in her concern. I still had the chance to recover the person I was losing. I was chasing a version of myself I emphatically walked away from, while she had already become someone else. I called to let her know I was coming.

My mind withered from the stress. My actions were driven by emotion. I disregarded everyone except her for the sake of temporary relief. I believed her support would help me face the uncertain future. I swallowed my pride and arrived outwardly composed. She made her way to my truck and the sight of her softened my heart. The world turned slower as my mind started to recover.

"Hey, is everything alright?" She asked.

I hugged her before violating the moment with conversation. "Let's drive around for a bit." I said, buying time to organize my response.

She cut the silence short with concern, "What did you need to talk about?"

I took a breath and poured out a heart I could no longer restrain. I voiced my fears and the difficulty in transition. It was liberating to release so much tension. Putting fear into words was freeing. I was feeling less isolated in a moment of vulnerability but not fully relieved. I was desperate for her attention and gave her the chance to respond.

"Well, you signed the contract." she sneered.

I made eye contact, frozen in disbelief with her statement. Her selfishness had no bounds.

"What?" I hesitated.

"You signed the contract." she repeated in a matter-of-fact tone and emotionally detached.

I had grown familiar with arrogance and challenged her pride with learned aggression. "Are you serious? Is that all you have to say?" Anger took precedence as I withdrew from an emotional collapse. I became defensive. "I can't believe you would hold that against me right now. I'm leaving for Iraq in a few weeks and all you care about is your own discomfort."

Before she could respond I caught a glimpse of the ring on her finger, defusing the hostility shielding me. I drifted outside my own body and lost sight of my reason for being there. My words fell on deaf ears with the violent image of myself made complete. Someone had solicited the promise of security and used my isolation against me. Her comment about me leaving her the previous year telegraphed an exit strategy all the while playing victim to my decision. My eulogy was notarized. The physical enemies ahead of me were frail in comparison to the emotional tyrant in front of me.

I drove home suffocating under a chaotic spiritual weight. I picked up Josh attempting to ride out the remaining weekend with a sense of dignity. We drove around the city taking in the scenery of a home we might never see again—a home I didn't much care to see. The casual conversations that usually lifted my spirit were helpless to pull me out from heartbreak. I buried my face into the steering wheel and turned the stereo up loud to drown the piercing cry of a dying spirit. I screamed with every cell in my body just shy of permanently damaging vocal chords.

My eyes blurred from the blunt force of emotion, tunnel vision, and dizziness. I coughed with a scratchy tone still unable to see with tears streaming. Josh froze in disbelief before taking control of the wheel. "Get out, I'm driving!" He forced the shifter to park before I could recover. My last bit of hope was stripped away as the walls of my stomach were lined with regret. My lust for her affection exposed a flaw in my faith.

Wills, powers of attorney, and finances were finalized over our remaining days in the States. I owned nothing of value at nineteen years old and watched all color evaporate from the world around me. I trudged through gear inspections, attempting to hide my heavy heart. I spoke only when spoken to, robotic and inconsolable. I said goodbye to my family in routine bearing and departed to leave my mark on history.

We marched down the tarmac toward an unmarked commercial airline with a lottery of one-way tickets. I recited my eulogy to an audience of one, feeling responsible for everything. I boarded with tunnel vision in deflated obedience. In less than 24 hours, the pearl white commercial bird would shed its feathers for an olive drab cage. Our final head count was an accusation of my presence–a presence I was being denied. Yet, the only Marine missing was Lira, who never came back.

CHAPTER THREE
CONDUCTING BUSINESS

JEWISH ARABS OCCUPIED RAMADI before the Second World War. The Nazi propaganda in Europe during the 1930's extended its corrupted agenda to the region of Mesopotamia. The Islamic domination of subsequent generations carried the torch and continued the anti-Sematic purge. On arrival, Ramadi was the largest city in Al Anbar Province and the focal point of the Sunni Triangle, home to roughly 400,000 Iraqis.

The insurgency retreated to Ramadi from Fallujah after heavy conflict with the United States the previous year. We would arrive in the prime of its volatility. My fears swelled and flushed between a crimson skyline and the Atlantic surface.

I was roused by nearby voices, peering into a dim cabin as static filled my ears. The subtle glow of laptops and aisle lights dilated my pupils into focus as I removed headphones to the steady hum of the jet engines. Most Marines were asleep in each of their economy-sized holding cells. The narrow seat on a half-world tour hardened my ability to sleep through masochistic circumstances. Eighteen hours of flight left us restless and numb to the remaining time of our commercial travel.

A brief landing in Iceland would satiate appeals and give our blood a chance to circulate. The M16-A4 rifle stashed under my seat hoarded valuable legroom. I lethargically adjusted my position to postpone muscle atrophy. For the next seven months, I was expected to carry my weapon everywhere as an extension of myself. There would be no forgiveness for any comfort gained at the expense of this faithful companion.

Our first stop in Europe was on the horizon. The trailing aesthetics of the States allowed my optimism to rebound with passive aggression. I posed a question to Burge, "So, we keep our rifles slung in the terminals too, right Lance Corporal?" My fatigued humor was rejected. "How about you stay on the plane doing push-ups while we get to stretch." He was not amused. The pilot interrupted with a brief weather update and unofficial travel notice:

"Good afternoon gentlemen, this is your captain speaking. We are on schedule for our arrival into the beautiful country of Iceland where the weather is sunny with a high of seventy-five degrees and the climate is as gorgeous as their women who are predominantly blonde hair, blue eyes and outnumber the men 2-to-1."

His comment broke the tension with appreciated unprofessionalism, sending the cabin into a roar of laughter. "That's what I'm talking about," Swanberg chimed in. The flight crew understood our sacrifice and met us on a human level. Some Marines reciprocated the social contract by explaining firearm safety and weapons function to flight attendants. They exchanged roles allowing Marines to give the seatbelt and life-vest safety briefs, even letting us serve food from carts.

We stretched our legs in Iceland and boarded again for Ireland where we were allotted a single alcoholic drink at the bar. My commitment to sobriety was challenged knowing this could be my last chance. Ireland restored our spirits with more than the approved amount Guinness. My cheerfulness quickly diminished at the thought of Kristi. Every conscious thought of her kept me searching for a distraction. I gave up my last ounce of control for a shot of Baileys Irish Liqueur.

KUWAIT

We arrived in Kuwait and dismounted in convalescent form. The alcohol eventually wore off, meeting the demands of a dry country. Equipment and checked baggage were staged on the flight line. The remaining evidence of home would migrate back west after refueling. All travel across the Middle East going forward was done under the concealment of night.

We marched off the tarmac into a staging area to receive ammunition. Warm summer air coated my lungs with parched dust of the flatland. Spotlights illuminated the crates of ammo surrounded by Hesco barriers and 7-ton trucks. The wash of floodlights sealed a perforated canopy and drowned the Kuwaiti horizon. Our presence was known in the light, but soon we would descend into darkness. Marines regained consciousness following our hasty dismount and resumed useless banter.

A few Marines unloaded their pretentious need to violate a sacred silence amongst an audience of lethargic bodies. A debate about superior rounds inevitably resulted as we were handed ammo.

"7.62mm is far more effective than 5.56mm." One Marine freely associated. "Have you ever seen an exit wound from a

7.62mm?"

"The 5.56mm NATO round wasn't designed for destruction," a fellow Marine spouted in a matter-of-fact tone. "It was designed to immobilize the enemy and allow the shooter to prey upon anyone who moves in to evacuate their fallen comrade, creating multiple casualties. You get more for your money!"

"Where did you hear that?" He responded with skepticism. "Did you just make that up?"

"No, that's what it was designed for. Look it up." He responded before yet another opinion emerged. "Well, an M-16 is more accurate than an AK-47 so it really doesn't matter how powerful it is if you can't hit your target."

Tepid entertainment postponed my emotional baggage. Sleep deprivation combined with emotional strain threatened me with complacency. Vigilance insulated my soul while I quietly thumbed 5.56mm rounds into my magazines. The jacketed green tip left me contemplating whether the steel core underneath was really designed for injury over death.

"So, would you rather be killed instantly by a 7.62mm or bleed out from a 5.56mm? Either way you die." They continued to draw out morbid scenarios. My mind faded into the concealment of night.

Our convoy shuttled us to a flight line of C-130's. The olive drab bird would guide us deeper into the boundless canopy of the Iraqi sky. The sight of military aircraft set a precedent. This was precisely what I desired and physically abhorred. *We have smaller rounds but better training.* I foraged for courage to feed on. *Smaller rounds mean higher capacity.* I denied the instinct of fear to commend myself on the desire for sacrifice. The tension called my faith into question again.

Hercules accommodated our formal arrival into Iraq, closer to the hostility I second guessed. A glimpse of the CH-47 Chinooks signaled the next evolution of our migration. A handful of Marines filled each tandem rotor-wing, enclosed by a tail gunner strapped to a .50 caliber machine gun. The blades pitched higher toward the canopy above the reach of small-arms. We penetrated hostile territory like a murder of crows silhouetted by stars. Vacancies of light under the perforated backdrop were all that signaled our position.

The glowing lights of the infamous capitol city resisted the heavy darkness, ominous and inviting. *This is really it.* My anticipation submitted in reverence. I contemplated life in Ramadi through the wonder of our mutual silence. My ignorance of Middle Eastern culture became obvious. *What was life like beyond the conflict?* The lights drew closer like a lion to its prey.

The silent bursts of machine gun fire broke my temporary buoyancy. I pulled toward the glass porthole to see tracers piercing the sky. The stream of automatic fire was narrated by the twin blades of our CH-47, masking all external noise. I glanced at the Marine next to me with a look of authentic surprise. The city substantiated its reputation before our feet could disrupt the Iraqi moon dust. The hostile introduction welcomed us with the etiquette of an esteemed adversary.

We touched down unscathed at Camp Ramadi. I marched from the back of the bird with my rifle at the ready. We stamped our boots in the powdery dust and revered the history we were writing in every step. Aggression stabilized my courage and veiled my fear. I would gladly uphold my contractual obligation to the lives around me. The political decisions that brought us here were muted in the snapping of rounds. No one fought for their beliefs in policy. The complexity of circumstances moved beyond theory. It was us versus them. The CO's words rehearsed in my mind. My juvenile convictions of the world were given a voice through the end of a barrel. The battle between heart and mind had begun.

RAMADI, IRAQ

Tent City was our temporary home at Camp Ramadi. The large Forward Operating Base (FOB) sat adjacent to Hurricane Point (HP) across the Habbaniyah Canal. We were given a few days to adjust before conducting our Relief in Place (RIP) with 1st Battalion 5th Marines (1/5). I glazed over the olive drab canopy covering rows of shoddy bunks, each lined with a strip of foam. The questionable racks were forgiven with the presence of air conditioners positioned along the walls. Downtime was the first challenge to overcome.

We were only allowed movement to the chow hall and back to maintain accountability. Weapons Company couldn't afford to lose track of its men before transport to HP. I closed my eyes again to accept the world I now belonged to. Finally, a chance to replace imagination with experience. I was forming of an identity through a role bigger than myself.

BOOM! My eyes darted open from the concussive wave and sound of an explosion. The first mortar burst just beyond our living area. "Incoming!" voices followed. We rushed to put on our flaks and dove under the impenetrable foam slivers. We remained helpless on the floor waiting out the destruction as steel rain concluded its volley.

I caught my breath from the abrupt demand for oxygen and scanned the room. Desperate prayers were sent up at random commencing a hasty revival of the unsaved. I didn't feel the need to pray, but found resolve in the company of new believers. Fear had finally revealed itself in the public arena even among a few

Senior Marines, exposing their humanity.

"All clear" rang through the tent–calling us out from concealment. I crawled out from beneath my rack curiously assessing the mixed reactions. The stress was exhilarating, though I was unsure how to respond. My heart was still pounding. Willy called out, "Sanderson, where you at?" "Over here, Lance Corporal." I responded. He looked blessed by the experience. The smile on his face said he thrived on the adrenaline. It was a welcome home ceremony. "Hell yeah, Sanderson!" he replied, feeling vindicated.

The threat of being killed this early on deployment was a buzzkill. The mental shift from fear to frustration converted me into a faithful servant to the call again. I was irritated that I might lose my chance to fight. I wanted to see the enemy. I wanted to know why Willy felt the way he did in the face of danger. I didn't want to go out lying on the ground. It was the first masochistic fear I had felt. I was afraid of missing out.

With only a few days in country, Battalion received its first casualty before our RIP with 1/5. Carson briefed the platoon the morning after roll call. "Listen up gents, a Marine from India Company was hit by small-arms fire. The round tore through his thigh piercing the Ka-Bar knife on his leg. This is the real thing now and what we've been training for. CAAT Teams from 1/5 should be arriving sometime in the next day or two to begin our transition. I want drivers and gunners to rotate out with their Marines when they get here. You'll be the first ones from Alpha Section on site at HP followed by an orientation patrol. For the rest of you that are staying here, remain in the tents unless you head to chow."

"Rah," we confirmed and broke formation.

The breath of demons brushed my neck, calling for

vengeance. The fight was on the other side of the wall. I didn't know if I was ready or if this would improve my faith. There was no choice in the matter now and only one way to find out. I buried my doubt and moved forward into the unknown trusting my heart would be stronger if I survived.

I awoke the following morning to another intimate encounter with the cold cement floor. The sound of gunfire and mortars cut our sleep short. The enemy's reveille was a call to arms. Vehicles from 1st Battalion 5th Marines arrived shortly after the attack. We stalled the convoy until our head count could be confirmed by the runner, but were forced to push when a delay threatened mission priority. I replaced the driver of the convoy's fourth vehicle, Willy took the gunner's position in the vehicle's turret. The rest of Alpha's drivers and gunners took their positions and pushed toward HP to link up with the rest of 1/5's CAAT section. Our first patrol would roll within the hour...

We stood ready to commence "business" as we staged next to the clearing barrels waiting for the Platoon Commander to give the mission brief. Anxiety loitered like an unwelcomed solicitor. I anticipated scenario after scenario of violent encounters attempting to reason my way out of damaging thoughts, but they were warranted.

If there was any side to the fight I wanted to be on, it was ours. We had armored vehicles with mounted machine guns, anti-armor rockets, air assets, and tanks at our discretion that could level any section of the city with copious amounts of collateral damage, all backed by Marines with enough anger to deny the enemy's humanity. I exhaled in relief knowing we were stronger and said a prayer with the prideful upper hand.

The platoon commander's delay became offensive. We

expected some kind of relief in the mission brief to ease the tension of our first patrol. His absence exceeded my ability to focus on mission details. In a matter of minutes, we would be driving the streets of the most dangerous city in Iraq. Waiting behind the front gate in a state of disillusion stole the valor of being a Marine. I died over and over in my own head waiting for the actual chance to be killed.

I caved to the raving anxiety. The thought of Kristi made matters worse. My sanity was slipping into hopeless despair. *Why did this sound like a good idea?* There were no time-outs, no pauses to reflect on my decision like before. This is exactly why I signed the contract. The idea of fighting was liberating, but the reality was costly. Courage moved beyond declarations demanding payment.

Lieutenant Awtry finally made his way to our huddle in front of the convoy. Destructive thoughts subsided upon receiving the mission brief. His voice cracked invidiously in the lowest key, "Alright Gents, bring it in." My expectation for a boisterous motivational dump was vetoed. Aggression evaporated in his tone:

"I just received word that three men were hit by a rocket that hit the chow hall at Camp Ramadi and two are in critical condition. Lance Corporal Swanberg was one of the casualties. We are awaiting more details but will have to wait until after this patrol. There are a lot of moving pieces at the moment, but we will revisit the event later.

Furthermore, our intent is to conduct a presence patrol integrated with 1/5 Marines to help familiarize you with the routes. VCs will guide us throughout the patrol, pointing out major areas of interest. Drivers: pay special attention to major streets and checkpoints. You will be expected to know the routes better than anyone. We will be moving quickly due to the threat of IEDs [Improvised Explosive Devices], small-arms, and RPGs [Rocket Propelled Grenades] so keep an eye out for wires running across roads and avoid driving directly over manholes. Stay alert and keep your head on a swivel."

We stood in shock, losing focus on the mission. Word of a casualty derailed my focus. His notification moments before patrol sapped the mental fortitude that held the shell of my being together. Relieved of all my strength, I was emotionally paralyzed.

"Mount up!" a senior Marine called out. We loaded our weapons and mounted the trucks waiting for the blessing from the Command Outpost Center (COC) to depart. I was frozen in disbelief and a passenger to my own motor control. The world around me ceased to exist. My heart stabilized simply from autonomic obligation. I struggled to keep from screaming into the sky. Every attempt to make sense of the situation failed. A voice interrupted through the radio, "Red Alpha you are cleared to push."

We proceeded out of the gate down main supply route (MSR) Michigan. The pressure against my lungs restricted the toxic air. My mind fixated on every possible threat as if each second in this city was going to be my last. The roar of our diesel engines were the only sounds anchoring my cognitive wandering. I faded down route Michigan through the vignette of watering eyes.

Movement throughout the city quickly liberated me from my hypnosis. People were going about their lives to and from work as if a foreign power hadn't occupied their country. The lack of concern for our presence delegitimized my fear. Iraqis walked the streets unfazed. The enemy I expected to face wasn't waiting for us at every corner. I hadn't considered what daily Iraqi life might really be like. Their casual demeanor recalibrated my perceptions and assumptions. My thoughts lapsed in time with a feeling of stability. If the Iraqis were just going to work then I was too. "Alpha 4 is up," the VC interrupted my thoughts with his Personal Role Radio (PRR) chatter to signal our position as the convoy's rear security.

"Copy, pick it up." The Section Leader responded from Alpha 1. I depressed the gas pedal and gained speed to maintain tactical dispersion from Alpha 3 around each turn. Nervously focused on the truck in front of me and overly concerned with the correct distance, the VC turned and addressed my abrupt adjustment.

"Keep it here, you're fine where you are. You don't want to be too close in case we have to back out of a tight space or one of us rolls through an IED. We don't want both trucks to get hit, or stuck in a potential ambush," he stated.

"Roger Corporal" I responded in an enthusiastic neurosis.

I was actively scanning for threats then caught sight of an Iraqi man walking his bicycle. The life surrounding us was a pleasant disruption from my own thoughts. The projected hostility in my mind moved from defense to confusion. Chaos signed its name on dotted lines across buildings and walls, but people were telling a different story. My mind eased at the sight of a casual Middle Eastern daily routine.

"That bridge up ahead is a pedestrian cross bridge so keep your eyes on civilians as you pass through," the VC warned. "Watch for people trying to throw objects onto the trucks. It's easy to toss grenades or Molotov's at our turrets." We proceeded under the first pedestrian bridge with a handful of Iraqis going about their day. Nothing in particular stood out but I heeded his warning. "You're more likely to catch observers scouting your movement." He continued to narrate our tour.

"OP [Observation Post] VA is up here on the right; the

Line Companies occupy these posts and conduct foot patrols in the area. Keep an ear out for radio chatter to catch their movement so you know where friendlies are." He added. To the left was a multi-level structure perforated with bullet holes. The corner of the top floor had collapsed, drawing our attention. The VC picked up on our suspicion.

"That is 7-story. We take fire from that building on occasion, if you can't tell." In addition to the collapsed roof, the building was damaged from multiple firefights and rockets. "Enemy combatants take shots at foot patrols and movement from there." He satiated our curiosity.

"What happened to the top of the building?" Willy chimed in. "A STA [Surveillance and Target Acquisition] team established a hide on the roof one night and was covering the area around OP-VA and Checkpoint 295. A group of insurgents infiltrated the building up to the floor beneath them. They set up explosives to collapse the floor from under them, but failed to fully collapse the level." the VC explained.

We cautiously approached Checkpoint 295 just beyond OP-VA. "Keep your eyes open for wires, bags, and piles of trash at major checkpoints. Vehicles are hit frequently here cause it's a high traffic area. We try to maintain a heavy presence to minimize their ability to plant IEDs," He spoke with a tone of acceptance. My nerves flared with curiosity. The threat of death cried wolf so often that fear had become an inconvenience to him. Iraqis driving on the same roads in proximity to our vehicles telegraphed the same attitude–hardly acknowledging the brevity of death.

We made it through 295 now holding behind the Government center in central Ramadi. I posted security on the southwest corner with other vehicles covering adjacent corners. We held our position for roughly twenty minutes, when a little girl about 8 years old peeked her head out from the gate directly in front of my truck. She jolted back in surprise when her eyes met the grill of our vehicle. A few seconds passed before she peeked her head out to see what we would do.

She cautiously peered around the door and met eyes with me. We watched each other in mutual astonishment. She was the first person to validate my existence in Iraq. I froze with rigid bearing, not risking an unfamiliar gesture. She lit up with a smile, grounding my spirit. I lost focus on the mission and everything that was building inside me. I couldn't help but smile back at her from my side of the universe.

I felt the purity of her ignorance. I wanted to see this city the way she did. She was an anomaly in the world I feared. Children like her were growing up in a city saturated with destruction. The regularity of our presence was an investment in their future. She was a light evaporating the clouds of my mental fog. Every minute in the city wasn't filled with firefights or explosions. We were at war with an ideology, not the Iraqi people. I was drawn into the culture in front of me unnoticed by a persistent vigilance.

The girl waved before running off. I sighed with foreign relief. The bitterness driving my insanity subsided. A few seconds of her presence changed something inside me. She wasn't angry. She didn't hate us. I had forged a polarized notion that Iraqis hated Americans. The young girl's extension of hospitality displaced my aggression. She saw the side of me that Kristi had given up on, the person I grew hesitant in stepping into. I was a foreigner to this country and myself, yet she saw me behind the ballistic eyes of an armored dragon.

Radio chatter picked up, requesting the section push out from the government center. We patrolled back to the MSR headed north of Checkpoint 295 into the Arches district. We navigated through residential complexes and traced a water treatment facility. Across Route Nova was a palm grove parallel to the Entry Control Point (ECP) for the North Bridge. The Euphrates River underneath impressed its eminence on me with historical reverence. The road trailing the Euphrates was a hot spot for IEDs, making the view one to be admired from a distance.

Pedestrian traffic was reduced to a handful of middle-aged males in thawbs. Kids ran through the residential streets in and out of houses as we passed. They seemed to be the only ones who paid attention to our patrols. Adult males briefly made eye contact and returned their gaze to the dust beneath their feet. My stress fell tremendously with our movement through Arches.

Soon enough, our patrol approached an area void of all life. "Keep your head on a swivel" a voice called from the backseat. A ghost town meant the enemy was near. The first two vehicles of our convoy rounded a residential corner as the third approached. Just then an IED detonated hammering our vehicle with a concussive blast. *BOOM!* The sound momentarily deafened unprotected ears while shunting motor function. I opened my eyes to a hellish portal of smoke and no leading vehicle.

I stalled our position in a suspended idle and risked separation from the convoy. "Hammer it!" the VC called out. I slammed the gas pedal and barreled through raining debris. All recently acquired peace was compromised as the jaws of fear swallowed us whole. I braced myself for impact into the back of Alpha 3 and disappeared in the dark grey billow of smoke and ash.

We emerged through the other side with a pull of fresh oxygen and a heart of fire. Alpha 3 was damaged but mobile. A call went through our green-gear announcing, "COC, this is CAAT Red Alpha, truck 3 is hit by an IED. Stand by for SITREP [Situation Report]." The transmission was followed by chatter on the platoon's PRR radios, "Alpha 3 this is Alpha 1, are you good?"

"This is Alpha 3, we're good. Push out of the immediate area and post security so we can inspect the damage. No injuries at this time. How copy?"

"Alpha 1 solid copy." The Section Leader in the lead vehicle radioed back to Weapons Company, "COC this is Red Alpha, Truck 3 is up, zero injuries at this time. We are posting security."

"Red Alpha, COC copy all." They confirmed.

"Hah, hell yeah!" Willy called out. We pushed out of the kill zone to a separate location and halted the convoy. I positioned my truck behind Alpha 3 to maintain rear security. Marines dismounted to assess the damage and determined it would drive back to HP on run-flats. The explosion rattled some feelings but failed to achieve physical harm.

My heart pulsed from the roof of my mouth, salivating iron. Excitement and fear fused together like a narcotic. The IED caused a release of endorphins tearing through my veins like the electric pulse of a defibrillator. I was alive again and addicted to a new drug with conflicting emotions. We survived the attack, inveigling me into a feeling of immortality. Alpha 1 called back to Weapons Company,

"COC, Red Alpha is up. We are RTB from Arches."

"COC copy."

We departed Arches down the MSR heading to Hurricane Point for our first debrief. We idled through the serpentine of the front gate, back into our place of safety. I was intoxicated with adrenaline and cortisol with an intervention of cultural assimilation. We staged at the clearing barrels to unload weapons. The sight of Hesco barriers on return reignited the words of our mission brief in vivid sobriety.

"Debrief in 5 mikes. Stage the vehicles and meet inside." A Marine from 1/5 corralled the platoon. We stepped inside to cover the nuances of our patrol. 1st Battalion 5th Marine's platoon sergeant addressed the Marines present, "Alright Red Alpha, let's recap on a few things…" I was far from being present, not really knowing who or where I was again. My attention slipped for a few minutes before tuning back in.

"It was likely a triggerman watching from a vantage point. The first two trucks made it through the kill zone while truck 3 took the impact, ruling out the possibility of a pressure plate. The bomb wasn't large enough to immobilize the vehicle but could have led to a follow-up attack with small-arms and RPGs. Your presence of mind to maintain bearing and push through to a safe area disrupted any follow up action by the insurgents."

My attention was divided. His voice was drowned out by pervasive thoughts of Swanberg. "For you 3/7 Marines, this is a hard start. I'll close here and turn it over to your Platoon Commander. Great work gents." He turned the floor over to Lieutenant Awtry:

"Like he said, good work out there. I don't want to beat a dead horse cause we're all still adjusting to the scene here, so I'll leave it at that." He paused briefly before addressing the stampede of elephants in the room. "In regards to the attack this morning, I'm receiving word that Swanberg was killed instantly and the others remain in critical condition."

He concealed his emotions while pausing to collect his bearing. He fought back tears and finished his comments, "We came in expecting a fight and that's exactly what we got. Take it easy for the rest of the day, I'll make sure we're not scheduled for any movement until tomorrow. That's all I have." He excused himself to recover privately, leaving us to resolve our broken grace.

In the trailing moments of our debrief, I sat on my rack for the first time to process everything that transpired over the last six hours. I wasn't immortal. I wasn't untouchable. I slipped back into shameful responsibility, alone again in the absence of home. Misery insulated my self-loathing in a room of shared grief. I faded into a state of burnout–my capacity for offense was shattered.

Sergeant Carson stepped in to address the few Marines from Alpha Section after the official debrief. "Good work out there today. I know we're all shaken up but this is what we came here for. There's plenty more waiting out there for us. We'll get our chance. I can promise you that." He looked around in a state of undefeated pride. It was his first display of approval with Alpha section.

"Don't bottle this up, talk to someone if you need to. Some of us have been here before, there's no shame in that." He retired.

I outsourced my sanity to the senior Marines, like a child mirroring an emotional response. They were stoics of a modern tribunal and vacant from previous deployments. They were calloused and I was newly afflicted. I was a foreigner being initiated into the tribe. Their indifference granted my citizenship. We became Marines in the communal reverence that stripped our humanity away. My silence resonated at their same frequency, authenticating an image earned in blood. Pride stood in the place of fear as shame was unleashing a fire in my heart. I finally embraced the searing violence defining my future.

CHAPTER FOUR
CONTROLLING CHAOS

THE SUN ROSE LIKE a charmed snake from each morning's call to prayer. Five prayers a day were broadcast from minarets across the city. Rural Iraqis traveling into the capitol city for work joined in reverence to their religious obligation. Some men wore jeans and t-shirts, others fought the heat in dishdashas' and sandals. More religious types wore Ghatras or Yishmaagh head coverings with an Egal. The color of their headwear symbolized which regions of the country they were from.

Iraqi women held far fewer freedoms than Americans. Women old enough to wear head coverings never made eye contact. Some wore only hijabs with blouses and jeans instead of niqabs, demonstrating some presence of Western influence. They were not allowed to drive and were required to sit in the back seat when being driven. We were advised not to confront them unless the situation dictated. Iraq was divided between Middle Eastern and Western influence.

46

With only two weeks in country, Weapons Company fully secured Hurricane Point. I approached Sergeant Carson fearing the 180 rounds of 5.56mm I carried weren't enough.

"Sergeant Carson, I only brought six mags. Are we going to issue out the spare mags left in the hooch?"

"How many firefights do you think you're getting into Sanderson? You gonna fight the war by yourself? You're fine with the ones you have, Rambo." He replied.

"Roger Sergeant." I said a bit confused.

The heavy guns mounted on our vehicles would see the majority of spent casings, but hadn't crossed my mind in the tunnel vision of eager fear. Our first duty on rotation was to man the ECP adjacent to the Euphrates river next to HP. The structure was used as a choke point to assess inbound and outbound traffic. The population grew exponentially during the day, as Ramadi was a hub for business. Drivers traveling into the city were directed into a staging area where vehicles could be searched for weapons and explosives.

The vacant brick and mortar 3-story structure we secured was anything but fashionable. The structure was hollowed out, with multiple outer walls missing on either side. It resembled a castle-like structure without the merlons and crenels. No parapets lined the stairs anywhere in the building and railing hadn't been installed on the upper floor. The building had no electricity and generators were brought in to power all communications and equipment. A COC was established on the main upper level, overlooking the Euphrates and entry for vehicle inspections. AT-4 rocket launchers and a Javelin missile system were kept next to machine gun posts in the COC.

The remaining rooms were fortified into defensive positions for Marines to watch over the surrounding neighborhood. Every other post had a mounted machine gun of either the M249 SAW

47

or M240B type. Some positions had grenades, smoke, and flash bangs while every Marine maintained their own personal rifle and ammunition. On average, rooms stretched 15 feet from floor to roof. One room on the outer edges of the structure oppressed anyone over 6 feet tall. The strange construction demonstrated a lack of detail in the building process.

We took our first posts, assessing positions for possible weak points. It was relatively safe from small-arms fire. A wall of sandbags rose eight feet in height with an isolated Humvee window resting in the center as a viewport. A chain link fence covered the gap between the ceiling and sandbags to prevent grenades from being lobbed over. Nothing about our position was aesthetically pleasing but was strategically advantageous.

Standing post was simply hours of baiting enemy fire. My vigilance would be tested. The ECP rotation was mundane, giving my mind hours to rot from hypothetical pollution. My new spirit conquered the monotony of introspection and wandered onto a spiritual battlefield.

I glazed over the eastern section of the Arches neighborhood toward the Warar district, conscious of my potential fate. *When is it going to be me?...* I focused on the landscape for half an hour before trailing off. *When can it be me? This heat is ridiculous.* The heat compromised my vigilance, reducing my mind further into a cynical trap. I hypothesized what kind of counseling I would get for being shot and disappointing the Marine Corps with damage to government property. *I am government property.* Boredom was driving recklessly.

Irrational thoughts were a strange confirmation of new mental episodes. Stress and fear were being undermined by cynicism. I was beginning to fear the conflict differently and conjured more complaints than worry. Complaints pacified anxiety, especially communal complaining—the descent led to diabolical thinking and twisted humor. Bitterness spoke over the worries in my mind. I sought refuge in the small victory of coffee runs from the Sergeant of the Guard (SOG) to disconnect from my own negative dialogue.

A woman wearing a niqab moved in and out of her home hanging laundry just below the viewport. A girl young enough to expose her hair followed to the backyard. Further down the road parallel to my line of sight, two men hovered behind the trunk of a car. Their posture shifted my attention. I raised my M-16A4 and scanned through the Acog optic for a better view. The red chevron swiveled their torso's hypnotically as I gauged their distance. *About 150 yards, easy shot.* Pointing a loaded weapon at a live target was a seamless transition now that bombs had been going off. This wasn't training. I insisted they brandish an AK type weapon or attempt to bury an IED to sanction my course of fire.

At that distance, the drop of a 5.56mm round was hardly an adjustment. The flight time was less than half a second and the fastest movement could be compensated with a slight lead. My excitement feasted on the momentum of our business culture, forming a reason to flip the safety catch and begin applying pressure. Eight pounds would cause the hammer to fall against the firing pin like a gavel. I sat anchored to the sandbags waiting for their conversation to become action. *What if they're unarmed? It could be nothing.* Rational voices intercepted my lead.

The heat turned frustration to anger. I needed positive identification to clear my conscience; I was committed to the act. *Pull out the RPG or shovel you have sitting in the trunk. Point your weapon in my direction.* Conflicting voices battled under presumptive recourse in the anticipation of a threat. My mind ran wild through unrestrained thoughts, brewed with boredom and stress. Cynicism evolved from complaints to aggression powered by a wounded mind. I was projecting onto them to deny a fear yet to be conquered. The two men drove off in their white Opel without affirming my hostility. I became the aggressor. *When did it become so free to think like this?* My faculties struggled to reconcile. It wasn't out of line to be prepared for violence. Projecting ill intentions on someone imposed death without trial. My untamed mind was deteriorating into darkness without discipline.

I didn't hate them but it made the job easier. Guilt was proof that my conscience wasn't fully seared. Training removed my personal identity in order to promote swift action in a fight. Aggression became a cudgel to enhance lethality, but dismissed remorse if it was not tempered with reason. I was uprooted in a moral struggle. Ethical decision making was reinforced throughout field training but the social climate was hostile. Going home alive was the number one priority and removing those in the way would be a burden of afterthought.

Even justified actions left Marines in a state of cognitive dissonance regardless of their beliefs. Humans were not built to war against each other and survival of this deployment became a spiritual battle. I was responsible for upholding my own moral convictions. We were not at war with Iraq, but an ideology that anyone could pledge allegiance to. Distinguishing between friend and foe was our profession. These men might have approached our location in need of food, water, or medical aid. If the situation had turned a dark corner, we would be writing a different story about Ramadi. *No better friend, no worse enemy*, I reminded myself.

Shooting a non-combatant was bad for business. We were there to eliminate insurgents, not create them. I tried to shed my guilt for presupposing a threat without positive ID. *What would other Marines do in this situation?* I thought. It was a violation of trust if one failed to act. I chocked it up to being overly prepared. I held the power to give or take life but doubt was killing me faster than the enemy in front of me. I felt compelled to embrace hostility as a safety measure and a cornerstone of my title.

The tension was hardening my heart. The conflict of overthinking subsided when I could successfully bury the voices of my past. I rationalized my way back to peace, instead found myself embracing the legacy I was thrusted into. I was searching for meaning in a state of unconventional joy. Clarity came through a Trijicon lens.

"If I were stranded without food, I'd eat another human to stay alive." Clifton declared. Conversations in the truck were unrestrained when boredom sought stimulation. Clifton would often sodcast his thoughts, keeping Alpha 4 hostage to a barrage of tangent scenarios. The night's dialogue landed on the topic of survival and the Donner Party. He continued, "In fact, I'd probably eat the thigh or butt cheek cause that's where the most meat is."

"So, if we were stranded somewhere and only our truck survived, you would eat one of us?" I baited him.

"Yeah," he said with pride. "I'd eat one of you."

"Are you trying to be funny Sanderson?" Clifton defended against our childish giggling.

Reis and Gonzales suppressed laughter from the back of the truck. Clifton's insecurities unraveled as he quickly perceived the humor to be at his expense. He would dish out back-handed comments yet was too fragile to receive them. Accepting my role

51

came with the courage to speak more freely.

The first week of ECP rolled into a week of night patrols. I cut a water bottle in half, filled it with coffee, and stashed it between the gearshift and radio mount. Planned distractions became a refined skill when conversations became dry. *Small victories.* Our movement fell under the cover of darkness and infrared headlights. Ramadi's urban guerrilla warfare made positive identification more difficult than identifying a uniformed military. Distinguishing between hostile and civilian was often impossible. However, the curfew imposed on the city isolated enemy movement at night.

Our current mission was to escort Task Force Iron Hawk, an Explosive Ordinance Disposal (EOD) team. They scoured the streets for IEDs to defuse, disarm, or destroy before having the chance to harm American patrols. EOD were angels in human form. IEDs in the urban environment were the biggest threat of the OIF theater and a form of ironic job security for these Marines. They would comb the streets of Ramadi with vehicles, robots, and their own hands to disable explosives. Their job was a blessing in country, but a cause for conflict at home. EOD was colloquially referred to as 'Every One's Divorced.'

Escorting EOD provided Weapons Company with the chance to learn the city at a snail's pace. Ramadi was new terrain and an urban battlefield. Getting to know the geography was a priority for a mechanized infantry patrol. Idling through numerous roads and intersections during IED sweeps imprinted a 3-dimensional map in my mind. Extended periods of waiting for clearance on a suspected IED before continuing movement dragged nights out. I prepared myself for a barrage of destructive memories and focused on our audible conversations in order to maintain sanity.

My anxiety diminished at the inconvenience of standing by

for indefinite periods of time. The unit preached on the dangers of complacency, adding to the mental war between trailing thoughts and maintaining ferocity. *Hurry up and wait.* Not knowing when a mission would end compounded frustration, further animating the stories.

We told stories about our lives before the Marine Corps to pass the dull hours of early morning. Most of us recently graduated from high school, while others completed a few years of university before dropping out to do something exciting. We laughed at those who left their comfortable life to sit in a truck and talk about being back in college. Patrols started with high expectations of an ambush or firefight, but often ended up with souvenir piss bottles and sexual exploits. Complacency was emerging through fatigue and a lack of enemy contact.

Kevlar and PVS-14B night-vision goggles (NVGs) caused massive headaches, pulling me back into my own discomfort. I adjusted my helmet every possible angle for relief but gave up in weary defeat. I rested my head against the window to support the off-balanced weight of my 14s. Before Clifton could finish his self-discovery monologue, the sound of a distant IED caught our attention.

We momentarily paused to assess the damage when gunfire interrupted from behind our vehicle. Willy jumped to his feet, keeping a low profile inside the turret. "I've got tracers directly to our six, coming from the east roughly 200 yards out." Clifton relayed the information over his PRR to all VCs in Alpha section.

A long burst from an RPK erupted no more than 100 feet behind the truck. I maintained a visual of the immediate area healed from all discomfort. Before our truck could take fire, Willy let out a burst from the .50 cal in the direction of the fire. Alpha 2 joined the fight with an M240B as a second volley of rounds screamed past us from the RPK. Willy sailed another burst from the .50 convincing the enemy to reconsider. The collateral damage from our heavy guns carried a reputation for ending firefights.

Iron Hawk pulled away from the exchange of fire. Alpha maneuvered each truck back into position to cover EOD and departed the area to break contact. Alpha 4 remained stationary with our engine revving in distress.

"Move the truck Sanderson, let's go!" Clifton demanded.

"Working on it." I replied. It wasn't moving. I began to worry that we'd been hit and the gear disabled. We became an isolated target the longer we sat there. I slammed the gas again. Nothing.

Clifton shot back again, "Sanderson if you don't get us out of here I'm going to…"

"Got it!" I interrupted him.

I glanced down at the .50 caliber shell casing blocking the shift from reaching gear. I brushed it out of the way, threw it into drive, and slammed on the gas. "Catch up to Alpha 3." Clifton calmed down. "Roger that." I said relieved. Radio chatter announced no further gunfire.

I scanned for secondary attacks through a haze of sensory overload. I caught a dark silhouette standing on balcony from a third story building. My nerves spiked. He had a visual on us long before we could assess his behavior. He could have been a trigger man from his vantage point. I questioned my hostile projections, remembering an undue stress from ECP. However, the attack moments before elevated suspicion. *There are still innocent people living here.* I maintained eyes on him until we cleared the area. His unfazed presence was an eerie indicator of insurgent collaboration.

We collected ourselves and made sure Alpha's head count

was up. My heart decelerated to a normal rhythm when 2 more IED's erupted. Another concussive wave struck us from less than 100 yards away. My pulse inflated again as debris rained down on our convoy. My judgment was compromised this time. I grew infuriated trying to make sense of the ominous presence on the balcony. The duration of our escort mission depleted all mental resources and left me fighting on reserves.

We finished escorting Iron Hawk and pushed back to Hurricane Point scanning roads and alleys for any last sign of movement. The break of contact returned us to a state of homeostasis and a taste for chaos. The stagnant hours of inaction were forgotten at the sound of gunfire. Conflict was the drug we thrived on. The combination of fear, adrenaline, cortisol and dopamine generated a deeper reverence for life. We were hunters in search of the most dangerous beast riding the line of being the hunted.

October marked our first full month in Ramadi. We braced for pre-holiday attacks before the official start of Ramadan, the holiest time of year for Muslims. In a few days our hope for a peaceful month would be tested. The holy days of Ramadan are spent fasting from dawn to dusk in recognition of the Quran being revealed to the prophet Muhammad. The Quran is believed by Muslims to hold divine laws establishing the five primary acts in Islam.

A season of fasting began the fourth of October. CAAT Red geared up for day patrols and security sweeps throughout the AO. "Infidels" were beyond the mercy of Islam and we were on the receiving end of their holy war. The peace we hoped to achieve was a distant cry as a single round snapped over Alpha 4 with no further incoming fire or complex attacks in pursuit. We pressed on with our patrol.

Moving south down Sunset Rd, we linked up with Kilo Company for heavy gun and vehicle support. Kilo Marines posted security on a suspected IED and requested our presence before EOD could arrive. We paralleled their security along Sunset. I maneuvered Alpha 4 near the center divider in the northbound lane to give Willy a better angle of fire.

Children roamed the residential streets within view, often approaching our trucks for candy. They rushed from their homes to beg for chocolate and soccer balls. Their panhandling irritated some of the Marines but reduced the likelihood of an attack. Being surrounded by Iraqis contested an enemy presence. We were learning to navigate around threats and fine tune vigilance for essential moments.

The heat of the desert drained moisture from every pore. We drank water beyond the limits of normal human consumption. Hydration was the key to our second biggest threat. Higher temperatures led to verbal warfare. The heatstroke forming in our mobile toasters turned Marines on themselves and anyone within eyeshot. "What's with the dudes wearing dresses? They think they're hard or something? What are they even wearing?" Doc Gonzales inquired. The Iraqi dishdashas looked like a dress to Western minds. "Where are these kid's parents?" He added.

Rip-It energy drinks pocketed from the chow hall and chewing tobacco compensated for the delusional banter caused by heat. Dip bottles littered the vehicles next to a few generous urine samples. I was beginning to appreciate the lulls in fighting during the dark hours of the night. Day patrols offered visual stimulation and a greater experience of the culture in Ramadi. Iraqis had an impressive sense of resilience with our presence in their homeland.

I dialed in on three kids casually perched on the sidewalk

just outside my window. I envied their place in the shade. They waved and motioned to their mouths, uttering the word "Chocolate" in broken English. The desire for chocolate was more important than their safety in proximity to a lethal threat. Either they knew and didn't care, or they didn't know and would have asked for more chocolate before taking shelter. Middle Eastern culture had me bound with confusion.

Just then a Kilo Marine opened my door to inform us that we were sitting a few feet from the suspected IED. "Well..." Clifton declared. "Back us out Sanderson," he said glaring down at the hole in the median to see what appeared to be a 155mm shell casing and wires. I backed up without hesitation and wondered if the heat displaced our vigilance. Clifton continued, "I know you don't like me Sanderson, but you don't always have to park the IEDs on my side of the truck!" *That wasn't the intent, but I'll keep that in mind* I thought.

"Copy that Corporal" I confirmed, partly to acknowledge our safety, the other to affirm his assumption. His attitude was unpredictable and became a problem for Alpha 4's morale. His comments were unwarranted though I forgive them under the pressure of fatigue and excessive heat. He had been staring into the hole without notifying the rest of us and blamed my driving to conceal his lack of attention to detail. At that point, I just wanted some chocolate too.

EOD arrived and immediately went to work. They placed a charge on the suspected IED and advised surrounding units of the countdown to detonation.

"2 minutes to det." A voice came over comms.

"Copy 2 mikes," the platoon sergeant confirmed.

We scanned our sectors for any suspicious activity that

might have grown with our time on site. Kilo Marines maintained cover behind buildings and jersey barriers while clearing the immediate area of bystanders. I looked back over my shoulder and saw my chocolate pandering friends chased off by a Marine.

"Thirty seconds," a voice cut through the radio–preceding a full count from ten. "...Three, two, one." the EOD tech relayed with a large *BOOM* at the zero marker. Debris hailed in front of our trucks littering the street with the hopes and dreams of killing Infidels. "Controlled det clear. IED confirmed, good eyes Gents!" EOD responded with an affirmation to Kilo's sharp eye. "Give us a few to pack up and we're Oscar Mike." "Solid Copy" We responded. We maintained security until EOD convoyed out. Iraqis in the area acted like rubberneckers in LA.

That evening, we smoked hookah and cigars at the fire pit, grumbling over the city's calls to prayer. I was opposed to smoking as much as I was toward alcohol, but thoughts of killing someone or being killed debunked the moral high ground. I smoked for the first time, giving up a straight-edge lifestyle. We were growing accustomed to the capitol city. The camaraderie developing during the platoon's downtime fostered relief, more settling than the nicotine available.

We established a routine by the fire pit to decompress after patrols. I played guitar to contribute something of value. Conversations between nearby Marines would cease to take in the sound of music. Firefights roared across the city like adjacent conversations. The separation between senior and junior Marines vanished in the flames of our congregation. Explosions from IED's and rockets made us spectators to a modern coliseum. Tracers and flares were fireworks from our global campaign. We listened in reverence, thankful for the times of peace.

The illuminating glow of flares silhouetted Ramadi's skyline revealing minarets and palm trees in a city under siege. We took turns identifying automatic fire and narrated the engagement. We could distinguish between the slower bursts of .50 calibers and 40mm to the cyclic 7.62mm and 5.56mm shoulder fired automatic weapons. Lengthy bursts from larger calibers meant Marines were laying the hate. Each one of us vicariously applied pressure to the butterflies. Our bonds were forged to the cadence of steel strings and cyclic fire.

I was fully alive in their company, resting with the first sensation of peace. I felt at home believing I could cheat death with the brothers at my side. We were invincible in each other's company. The sound of firefights confirmed our strength and deceivingly silenced fear. We knew every day could be our last but the longer we lived the less we believed it. I thought of the 1/5 Marine during my first patrol and the inconvenience of stress. Fear naturally subsided, though it never fully disappeared. It was an unavoidable tax in the business of killing.

The night carried on with the faint cracking of rounds and pallet wood. The soft concussion of what sounded like an IED startled our peace. Our concern was communicated through mutual glances. Emotional weight vented itself in large sighs as we counted our blessings and conserved our energy for the next day's fight.

*   *   *

The following morning's sunrise streamed the sky with red and orange banners held up by the smolder of spent tobacco and pallet wood. Camp Blue Diamond across the Euphrates could be seen parading a palace once belonging to Saddam Hussein.

Rumors went around that snakes were found along with various equipment in some rooms signaling that people had been tortured in the past. The military presence on the western end of the city awaited threats from the east, where light emanated beyond the horizon–forewarning of the storms heading our way.

PFC Christopher Mallonee, a Marine from Arkansas, and I made our way back to the hooch after the morning chow passing the Battalion Chaplain who greeted us in his casual officer tone.

"Good morning Sir." I called out comfortably.

"Hey, morning guys. How are you two holding up?" he replied.

"Well you know, living the dream." Mallonee sugarcoated. The Chaplain forced a laugh under his breath, not in the usual cheerfulness we anticipated. "We were headed to the phone center but it's shut down for the next 24 hours," I tried to cover up the awkward pause. "Battalion had another incident but we haven't received any details." I concluded, realizing that the Chaplain already knew.

"Have you heard anything?" Mallonee spoke before thinking.

The Chaplain's brief pause was brooding and revealed the price of ill-considered small talk. "Actually, I have." He responded with caution. "I don't want to speak too soon, but Lima Company was hit by an IED and reported 1 KIA [Killed in Action]. I'm still waiting for confirmation but we're almost certain it was PFC Bedard."

The stab of denial pierced my ears as I stood defenseless at his reply. "I'm preparing another memorial service in the next few days…" He went on. I checked out at the words "Waiting for confirmation" embracing all denial.

"Bedard?!" Mallonee interrupted. "Like, Andrew Bedard?"

We looked at each other in disbelief before returning our

attention to the Chaplain.

"Yes, did you know him?!" He noticed our suspended belief.

"Um, yeah. I mean yes sir. He used to be with CAAT Red before he transferred to Lima Company." I said losing formality.

"Oh, I'm really sorry to tell you guys like this. I didn't know he was a friend of yours. If you could do me a favor and keep it between us until they confirm everything." He concluded.

"Sure" I closed and we parted ways.

Anger rapidly diffused a swelling fear. We were not invincible even with the courage of lions. Guilt filled my heart wondering if it should have been me. Denial was invalidated through the mental fog. I departed knowing everyone in CAAT Red would entertain the same thought. Bedard's death would not be received lightly.

LCpl Jeffrey and a few others were asleep when I returned to the hooch. I felt it was appropriate to break the news from within our peer group. I couldn't bring myself to remain silent then acted shocked when the news reached the rest of the platoon. I trudged through mental intrusions as if rationalizing the situation could change the grieving process. I gathered my mind and stood beside Jeffrey's rack for a moment before nudging him awake.

"Hey man, can you wake up for a min? I've gotta talk to you."

He rolled over and took a deep breath while trying to force his eyes open. A few more disgruntled breaths were let out, "What's going on?"

"I don't know how to say this but I figured you would want to hear this from within the platoon. I just talked to the Chaplain..." I paused momentarily.

"It's about Lima Company's engagement last night. Bedard

61

was on that patrol and…well, the Chaplain said he's preparing a memorial service for him."

I stood in desperation waiting for his response, instantly regretting the manner in which I presented the news. I felt responsible for his death. I knew he did too. The shock carried a paralyzing intensity and the silence was deafening.

The cycle of disbelief telegraphed through his body language. His eyes were wide in response. I kept quiet to avoid useless chatter. Our internal worlds collapsed while maintaining outward bearing. We sat there in a cycle of failed rationalization. He struggled to inhale again. I felt the emotional weight sinking as he dropped his head to consider my words more carefully.

He finally broke with a look of mirrored doubt, "How do you know?" He said, despondent in his response.

"I just spoke to the Chaplain. He's been in contact with Lima and said the funeral service is being arranged later this week at Snake Pit."

"Alright, thanks." He closed with a dissenting nod. His eager skepticism was rooted in the same painful hope that I held onto. He swallowed hard to keep from falling apart. We each pulled a heavy breath of seemingly unfiltered nicotine in the absence of oxygen. We silently suffered remembering why we bleed. As word spread through CAAT Red tensions reached their breaking point.

I awoke to shouting at the back of the hooch in the morning. "Get outside, we're gonna settle this right now!" Clifton snapped at Jeffrey as he removed the rank from his collar. The quarrel gained the platoon's attention as they proceeded outside to handle their business. It was the first public act of defiance between a senior and junior Marine. Jeffrey lost his emotional bearing when Muniz ordered all junior Marines to wear full gear during their night watch. He vehemently rejected the disrespect on a day filled with grief and pushed back.

Feelings were irrelevant once rank was involved. Clifton capitalized on the opportunity to exert power over a subordinate rank but it backfired. His pride was damaged when Jeffrey dislocated his shoulder in their scuffle. Sergeant Carson ordered all junior Marines outside as a result of the quarrel. Jeffrey was ordered to fill 1,000 sandbags without the help of anyone else. Disorder was never tolerated. The loss of a brother couldn't detract from institutional hierarchy.

I disregarded the order and helped Jeffrey fill sandbags. One by one other junior Marines did the same. Everyone recognized the unspoken fight for dignity. We were all afflicted by the same mess but were not afforded bereavement. Carson was no stranger to loss and perceived the tension. The expectation to quietly bury trauma for the sake of order was a masochistic fallacy. I buried my feelings in the sand with the rest of the junior Marines, filling sandbags with inaudible contempt and a quiet appreciation for life.

Character was revealed in the greatest moments of pain. The loss of Bedard fractured the platoon's morale. I failed to understand the weight of Carson' own burden and responsibility he may have felt for Bedard's life. Surprisingly, senior Marines came out to help fill sandbags. The breach of hierarchy from their side meant a human barrier had been broken. Carson joined in as well, bringing recognition to the limitations of order. In quiet submission we all became human again.

The combination of fear and frustration subsided after filling sandbags. I still struggled with anger toward the Iraqis when painful vices arose. It was easy to blanketly accuse them to justify my pain. They were an easy target after killing a friend. Aggression was contagious, though oftentimes rooted in suffering. Anger was useful in combatting fear but gave rise to a new demon at the cost of empathy and remorse. Violence was the cost of doing business. Vengeance offered an emotional escape when all attempts to rationalize the circumstances failed.

The conflict in Ramadi was a dynamic evolution. Battalion briefings kept Weapons Company up to speed regarding insurgent activity. Al-Qaeda in the Anbar province was comprised of a militia–a collection of men from surrounding countries. There were Chechen fighters from Russia and a Syrian sniper among locally recruited Iraqis. Some insurgents consisted of Egyptians, Saudi Arabians, Jordanians, and Iranians. The majority of the fighters were Iraqi Nationalists. Groups such as the Ba'athists and Sunni and Shia militias were mostly native to Iraq. We were not at war against a definitive enemy, but the intentions of men's hearts, shifting by the day.

Support for the insurgency declined, which gave rise to extreme measures. Insurgent groups recruited local men to take up arms against U.S. forces. Uninvolved parties were forced to fight against their will. Locals were now placing IED's and shooting at military convoys to keep their families from being killed. Money, food, and life were the rewards for supporting Al-Qaeda and psychologically conditioned civilians into submission. Filming attacks became the ticket to freedom and generally took place from positions of visual advantage. Spectators became suspects signaling hostile intent.

The Iraqi Government sought to eradicate Al-Qaeda's presence and regain control of the AO. Local support would be required, yet stood conflicted. Some IED's were thrown hastily on open streets without any attempt of secrecy resulting in botched ambushes. The ineffectiveness of hasty attacks was proof of inexperienced fighters and civilian recruits. Small-arms fire and IEDs varied in performance. Explosives large enough to cause damage required time to be buried or disguised. The damage we created outweighed the enemy's and we were careful to avoid collateral damage when possible.

Intel came through regarding a group of fighters who may have coordinated the attack on Camp Ramadi around the time we arrived. Rumors scoured the platoon that these might be the men responsible for Weapons Company's first KIA. *Swanberg*. This was a chance to achieve justice. Rumor or not it was all we needed to get our minds back to work.

We executed a raid on a residential compound detaining three men. They were blindfolded, flex-cuffed, then loaded into our vehicles. The high-back held two detainees and two Marines from our section. The third detainee was placed in the rear seat of my truck. He sat shamefully with his head buried in his chest. Marines were convinced these were in fact the men responsible. We accepted it to substantiate a hatred eating away at our souls. We needed to feel a sense of justice even without verification.

Discussing how to bring them back in body bags filled every void in our hearts squeezing the rest of our humanity through our pores. The language barrier was all that shielded their dignity. We wanted these men dead. They would face a firm sentence if convicted. *If.* The bureaucracy of politics might release them in the absence of evidence. Turning them in dead was the closure we wanted. I channeled my anger toward the scapegoat in the backseat. The Marines we lost, the sandbags that were filled, the woman who left me, it was all coming together. He was at fault and very little stood between us.

I dismounted the truck at Hurricane Point to clear my weapon and exhaled with pride at our catch. I peered through the window at the blindfolded man in the back of my truck awaiting further instructions. Just him and I, in a moment of silent contemplation. I imagined what might be going through his head. His posture was a statement of forfeit, a submission to defeat. My eyes fell hard on myself, exposing the demon casting judgment. *Was he a dedicated insurgent or just taking up arms to avoid having his family killed?*

I revoked judgment in the clamor of possibilities. I wanted these men out of the fight and they were. Their heads hung with shame but failed to sway the social verdict. I hated the evil that dictated their actions. I hated the incentive that drove men to kill in support of a corrupt agenda. In their eyes we were the same, we just had the upper hand. The difference between good and bad was impossible to determine without a uniform. Yet even with a uniform, no one could perceive the heart of either side.

CHAPTER FIVE
ADAPT AND OVERCOME

THE SUN ILLUMINATED THE FACE of the Channel Islands, brushing the Pacific coast to the Point at C-Street. Cory, Josh, and I set out early in the morning to surf our favorite spot just north of the pier. Cory fixated on a dawn arrival. Waking up early on a weekend in order to have a decent surf session was well worth the lost sleep. It wasn't an inconvenience knowing we were in control of the agenda.

"We're probably the only ones who get up this early on the weekend." I directed my frustration at Cory.

"Dude, you gotta be there for high tide man otherwise you'll get skunked. You're not gonna find parking if you get there late fool." he replied eager to get in the water.

"We need to stop for coffee anyway. And I need a smoke." Josh added.

"Ugh, you friggin coffee drinkers." Cory jabbed impatiently.

"Just make some, it'll save time." I said trying to appease his

67

impatience.

"He doesn't have a coffee maker." Josh added with sarcasm. "Mormons don't drink coffee, right Cory?"

"What, you don't drink coffee?" I asked.

Cory replied before heading out to pack up boards and wetsuits, "Nah fool, I just don't like it. Let's go dudes, so we can get your stupid coffee."

Our first dive into glassy water washed our spirits clean from the weekly training regimen. The stillness of the water was pierced by neoprene and polyurethane, along with the occasional blow of air from nearby dolphins, splashing water like tiny depth charges. The Pacific Ocean flooded our wetsuits with chilly temperatures far removed from the high desert. I surfaced from my initial submersion, pulling a deep breath of air from thermal shock, transporting into a timeless world perfectly at peace with its own order. We closed our distance with the approaching sets to reach the glassy flat beyond their crashing curls. The fight against the tide was worth the freedom that came while searching for that first ride.

I opened my eyes from the brief reflection of home, wishing I was waking up at 6:00am to surf. The bond of trust within our vehicle team was disrupted when a logistical decision moved Willy to Bravo Section without warning. PFC John Schaefle from Illinois, became Alpha 4's gunner. Individual rotations within a platoon were rare, but were utilized when Section Leaders could mitigate issues and maximize team potential. Full rotations through day patrols, night patrols, QRF, and ECP were familiar to all of us now and the shift was a minor adjustment.

Two months in country hardened our spirits and built

confidence in our movement. Our Humvees were branded from gunfire and IEDs. Daily calls to prayer began to sound like the rally of a wolf pack. The final days of Ramadan were testing the resilience of Weapons Company. Alpha Section received another screening mission for Kilo Company. Our patrol led us to the Jamhori district to cover Marines on foot. We split the section in two: Alpha 1 and 2 headed south with Alpha 3 and 4 pushing east. I scanned streets maneuvering around corners through narrow underdeveloped roads. Bravo Section maintained security to the east.

Soon enough, we heard an explosion to the west and scanned the immediate area before requesting a SITREP. A call for QRF came from Alpha 2. We set out to link up with the rest of our section bearing in the direction of the explosion. Comms with Alpha 1 were severed.

"Alpha 1, this is Alpha 3." No response.

"Alpha 2, Alpha 3, what's your SITREP?"

"Alpha 3, this is 2. Truck 1 is down on Sunset adjacent to Al Huz. They are unable to move. We are holding security. What's your ETA?"

"In route, two Mikes." Alpha 3 reported. We slammed the gas pedals to the floor with diesel engines screaming. Alpha 2 covered the immediate area with the .50 cal, stranded without us.

Alpha 1 hit a pressure plate rigged to a massive IED. The bomb detonated with enough force to flip a Bradley. The explosion lifted the Humvee from the pavement, slamming into earth without a front end. Everything in front of the windshield was missing except for the smoldering remains of an engine block. Their slow movement saved the truck from significant damage had it detonated directly underneath.

The concussion knocked everyone unconscious and presumably dead. One by one they slowly regained consciousness. The VC panicked at the sensation of fluid soaking his cammis. He patted himself down relieved to discover anti-freeze instead of blood. Other Marines came-to, instinctively collecting equipment. They dismounted the M240B weapon system and raided a nearby house for cover. They were secured indoors by the time we posted security on the block.

My heart stalled at the sight of Alpha 1 beyond recovery. The Platoon Sergeant hustled our truck to relay information due to the severance of radio communication.

"What's the status?" Clifton said.

"We're standing by for a flatbed truck to evacuate the vehicle. We secured a house across that road. Headcount is good, just shaken up. Sit tight here and cover the road until we can push out." Staff Sergeant Yellope responded in haste. He bounded toward Alpha 3 after giving us the update.

SNAP! A single round ricocheted off the ground missing Yellope's foot by mere inches.

"Schaefle, did you see where that came from?" Clifton called out.

"No, but I heard it. It's a good distance beyond the road behind us." He responded.

Schaefle scanned for movement and I searched my sector of the road but was blind to the direction of fire behind the vehicle. The shot likely came from a sniper set up at a distance to engage Alpha 1 after the initial ambush. *This was a coordinated attack and we're in the kill zone.*

We continued scanning for threats as Yellope took cover. Schaefle's blood boiled at the chance to unleash the .50 cal. Vengeance was the salt salivating our tongues. The lack of positive identification made engagement difficult. Insurgents fought with guerrilla tactics because they were outgunned. Frustration clouded the ethic of collateral damage. Gunners just wanted to engage. CAAT Black arrived as QRF to reinforce our patrol in our enraged state of vigilance.

Marines from Alpha 1 loaded into an armored high-back and evacuated through the backside of the Government Center en route to Hurricane Point. Our patrol came to a halt when the stream of an RPG sailed from behind my truck. I braced for impact. The rocket missed Alpha 4 and impacted a two-story house to our right. Small-arms fire followed the explosion. Multiple rounds of 7.62mm peppered the road, walls, and back end of our vehicle. Clifton and I turned to Schaefle for an observation. He was our only set of eyes from the turret.

"Can you see them?" Clifton shouted. "I've got nothing!" he responded, "There's too much dust, I think they bailed." Marines from Alpha 1 being transported in the high-back stood over the armored walls returning fire with their M-4s. They weren't going to miss the chance to exchange rounds after nearly being killed by an IED.

I spectated the exchange of fire from the driver seat blind to what was behind us. We still weren't moving. *What is the holdup?* I posed a rhetorical question to myself knowing the mutual thought would go unanswered. Only after the firefight ceased, we made our way past the Government Center. "Alpha 4 is up," Clifton called out on Route Michigan.

My blood had finally re-oxygenated as we pulled through the gate at HP, when a volley of mortars sent us scrambling for cover. Once everything died down we laughed off the cardio-vascular strain. Cynical humor was the only way to cope. I dropped my flak as Schaefle remarked, "Just another day in paradise."

A few days after the catastrophic IED that crippled Alpha 1, CAAT Red was back on the streets looking for blood. We executed an early morning raid on a house in the Arches; another "presence patrol" to fatigue our mental focus. The first three trucks rounded a corner with Alpha 4 in pursuit. Shortly before completing the right turn, my attention was drawn to the rear before briefly losing consciousness.

BOOM! An IED tore into the passenger side of our Humvee. The explosion dampened my hearing to a steady hum. I ducked and turned my face away from an immense heat. I looked to my right to see if Clifton was conscious when a second explosion consumed the truck. I buried my face in my flak and lost sight of everything to a cloud of black smoke. Concussion stripped our vigilance like a rug from under our feet.

The dark cloud poured through the gunner's turret stealing our breath and bled tears from our eyes. I coughed the aerating poison from my lungs, fanning the smoke and ash from my line of sight. I regained visual of the road through watering eyes and slammed the gas pedal to evacuate the kill zone. I glanced to my right again and called out to see who was conscious, failing to realize they had temporarily lost their hearing as well. Reis's Kevlar was on fire and quickly patted out by Doc Gonzales.

The ringing in my ears placed my attention back on the road in the absence of audible distractions. I coughed the remaining smoke out of my lungs trying to gain sight of Alpha 3.

Clifton called over the comm, "Alpha 4 is hit..." *Cough* "...Keep pushing."

Alpha 3 came back, "Roger, push ahead of us we'll take rear security."

"Alpha 4 copy," he replied.

We came to a stop a few hundred meters past the blast site and positioned our trucks to rig for tow. The blast blew open the trunk exposing us to potential small-arms fire through the blast shield port. We rigged the truck for transport with Alpha 3 taking rear security. I glanced down at my fuel gage—the needle was now on empty. We were leaking diesel and just shy of engine failure. We towed Alpha 4 to HP in haste.

I inspected the vehicle's damage at Motor-T, running my fingers across the scars etched in the armored cabin. Shrapnel carved the doors like comet trails exposing steel beneath tan skin. Windows were splattered with spider cracks, resembling dozens of webs spun to ensnare prey. The rear tire well revealed a hole large enough to fit a softball displaying the large piece of shrapnel's path of travel. The baseball-sized object tore through a case of MREs and into the fiberglass body of the AT-4 rocket. The second explosion was likely a result of the AT-4's charge and the reason why the blast shield and trunk were exposed. An actual discharge from the rocket would have carried us back to the States.

Alpha Section's debrief believed the IED was set off by a triggerman since the last vehicle was specifically targeted. Pressure plates detonated on contact with the first vehicle and timers would have been risky to be that effective on the tail-end of a convoy. We knew they were using spotters but no suspects flagged our attention during the patrol. The attack was beyond our control but we were alive and well. The feeling of invincibility was revoked in a state of fearful wonder, though a deep sense of gratefulness replaced cynicism.

Hands on the clock spun like a roulette wheel, insidiously strafing numerals as if they stood for each of our names. The five months left in Ramadi harnessed my attention away from the blast that nearly erased Alpha 4. Over the past three days, Weapons Company had suffered two IED attacks in a single section. The threat of harm to any Marine was enough to withhold our empathy for non-military losses. It was a cost we all agreed to and another rise of the sun was good for business.

CAAT Red carried on with patrols collecting emotional debt. An Iraqi woman signaled to our convoy that she needed medical attention. She cried out to us in violation of cultural norms on approach. A bullet tore through her leg and she was attempting to stop the bleeding when she signaled Alpha section for help. Gonzales insisted we stop to help but was denied in the name of mission priority. She was likely wounded as a result of tribal feuds. We pushed on.

The priorities of "mission accomplishment" saw through her like a piece of broken glass. Her screams were softer than the ringing of my ears. The strict adherence to orders under-mined the hearts hidden behind digital print. It wasn't for a lack of concern–tactical decisions were above our pay grade. Our perceived moral bankruptcy was sure to employ EOD again.

The title I was earning was far removed from fancy billboards and motivational recruiting posters. The utilitarian nature of the Marine Corps was turning a profit so successful, it could take a life before the experience of death. Our personal honor was on the line when low intensity missions held priority over crucial aid. Not every moral battle could be won but we were dominating in the war. I was grieved beyond composure as her voice faded behind the exhaust of our vehicle.

Orders from higher were making significant strides for the future of Iraq. The future of Ramadi was in our hands and ultimately a business transaction. It was becoming routine to hold meet and greet missions with local religious leaders. Seeing the local population suffer because of our presence was being addressed even if it didn't make sense on the ground level. Platoon commanders and above coordinated tea dates with tribal leaders and Sheiks to find mutual solutions for peace.

"Alpha, stage trucks at the clearing barrels in five mikes!" Carson announced stepping back into the hooch from a Commander's brief. Weapons Company received intel on the location of a local IED maker. Alpha Section departed Hurricane Point at 0400hrs to execute a raid on a residential compound and secure the block with 360-degree vehicle coverage. Marines dismounted under the cover of the night and made entry. Intel led to the arrest of another insurgent. Family members wailed in protest but were silenced by force. Once secured, we made our RTB and awaited further orders.

Radio traffic picked up as India Company began taking heavy fire. Alpha section remained staged at the gate in the event QRF was called out. Word of our detainee's capture sparked a hostile response and resulted in an immediate counterattack on American forces. India was able to break contact with the enemy without QRF but sustained casualties. One Marine was undergoing surgery, and another was confirmed to be KIA.

The end of October brought CAAT Red back to ECP vetting inbound and outbound traffic. I made two arrests and placed the detainees in a holding cell. Iraqi Security Forces (ISF) working alongside us claimed that one of the individuals was a suspected enemy sniper and killed one of their family members. Not long after their detention, a Human Exploitation Team (HET) arrived and escorted the detainees to a secured room never to be seen again.

In recent weeks, an Army convoy hit two substantially large IED's. The explosions flipped two Bradley tanks, leaving soldiers inside vulnerable to an ambush. They were hit by small-arms fire as they tried to crawl from the burning wreckage. Those who remained inside to avoid rifle fire were burned alive.

Marines standing post in proximity to the attack attempted to repel fighters from angles of opportunity but were vastly restricted with friendly forces down range. The escalation of enemy attacks meant we were doing damage to their command structure.

Marines from Kilo Company saw their chance to avenge our loses while manning a machine gun post near Arches a few nights after the Bradley attack. Two military aged males (MAMs) were spotted dropping bags along the roadside, which signaled the placement of an IED. A Marine spotted their movement and engaged with a burst of 40mm rounds from the MK-19. The short burst decimated the area and the men disappeared in a cloud of smoke. A battle damage assessment (BDA) came through the radio after the smoke cleared.

"Whiskey COC, this is Kilo OP. We have engaged 2 hostiles in the Arches area, one possible KIA. Requesting patrol with high-back to assess the scene, how copy?"

"Kilo this is Whiskey COC, confirm you have one KIA and need transport?"

"Affirm. Be advised, one wounded MAM crawled out of our line of sight." Kilo responded. "Solid copy, rolling CAAT team." Alpha Section received the order and awaited the mission brief.

"Alright gents," Yellope called out to the VCs huddled at the front of the vehicles, "Kilo engaged 2 MAMs in Arches with one possible KIA and another wounded and missing. We will secure the location and assess the damage. Once secure, we will load any bodies in the high-back and return to HP."

Alpha was cleared to push and rolled from Hurricane Point with NVGs and infrared headlights. We secured the location without retaliatory attacks and dismounted to inspect the immediate area.

One lifeless body and a bloody trail from another were discovered. The casualty was able to escape but likely died not long after. Kilo's BDA was accurate. The high-back pulled up to collect the deceased.

Alpha 3 radioed, "Alpha 4, pull your truck behind the high-back and give us some light." I pulled our vehicle into position. Dismounts recovered the body minus the head and searched the area for weapons. Kilo's MK-19 round made direct contact. I focused my PVS-14s as they loaded the body onto the litter. The high-back doors swung open and the stretcher was laid in the bed for transport. The lime-green glow of my single monocular masked the color of spilled crimson like the vile accumulating in my stomach. We mounted up and pushed back to HP.

The high-back pulled off to an area designated for human remains. The man hadn't been dead long enough to dispel all fluids. Blood pooled on the litter mixed with fragmented brain matter. The volatility in removal was amplified as Penney lifted his end of the stretcher too soon. Lance Corporal Harkins, a fellow Mortarman from California, was on the receiving end. He stormed off in disgust after being drenched in warm fluids–leaving me with the cleanup. I scrubbed the remaining blood from the canvas and floor board of the high-back trying to avoid physical contact with grey matter.

My own heart turned against me under a crushed spirit. I was angry, feeling exploited–left to heal in my own time. I buried my vices to avoid crippling others and carried on as if each problem would be tomorrow's battle. I was losing the voice of reason in the echo of after-action reports and recurring mission briefs.

The primitive nature of my being felt victorious from emotional collapse. The thought of neutralizing a threat was empowering in the deepest way. We fed on the smell of carbon and smoke. Fires within the human heart burned hotter than the summer temperatures. There was something intimately liberating about the freedom to ensure survival when fearing death. Our hatred for the enemy was a cornerstone for the preservation of life and illuminated by the love we had for each other.

Ramadi became mysteriously familiar. The singularity of our professional role eliminated the barrage of endless responsibilities at home. I reflected on recent memories and how driven I was to arrive. The fire that brought me here was the flame consuming me. I was wounded by my past decisions for investing in a painful future yet forever grateful for the veil that was being removed. Sacrifice promised the vanity of glory but was delivered through an understanding of life in the community of broken souls. We shared our scars in exchange for peace. Flames were cauterizing wounds and callousing layers of our hearts.

CHAPTER SIX
EXCHANGING FIRE

THE SUMMER HEAT RECEDED, ushering in a wave of cooler weather like an answer to long forgotten prayers. The sun was pinned high over Ramadi while families at home gazed toward the reflection of our light off the moon—at war with their shadows. Choosing to be grateful was a chore in the context of our profession. CAAT Red returned to QRF per rotation. Bravo section was primary QRF with additional support from Alpha. A call for QRF went out around 1700hrs and both sections were dispatched to support an EOD team hit by multiple IEDs.

They were defusing the initial IED while unaware of a secondary underneath their position. Both bombs detonated with catastrophic effects. Bravo was the first on site and began casualty evacuation (CASEVAC) procedures by the time Alpha section arrived. They maneuvered into position, posted security, and dismounted to assess the blast site. Injured Marines were rushed to Charlie-MED as remaining trucks stayed on site to repel tertiary attacks during clean up.

The smoldering wreck of EOD's vehicle required a flatbed truck to extract. Marines from Bravo section scavenged through the downed Marines' gear covered in blood. All weapons, equipment, radios, and tools would need to be washed from the fluids they were stained in. Some Marines would return home to rest on sacred ground.

Small-arms fire surrounded our position as both sections maneuvered to evacuate the wounded Marines. Alpha screened to counter the ambush, realizing the insurgency was a step ahead of each movement. Bravo finished clearing the blast site while under fire. Every truck with a line of sight exchanged machine gun fire. Tracers and 7.62mm rounds swarmed the air like wasps. All of CAAT Red's aggression could be felt through the growl of the M240s, MK-19s, and .50 cals. Marines on the ground augmented the fire with M-16s.

Both sections debriefed to gather perspective on the attack. Two Marines were killed instantly from the blast and a few others were wounded. One of the wounded Marines succumbed to his wounds at Charlie-MED. CAAT Red arrived with enough time to prevent further casualties. EOD was acquainted with catastrophes like this, yet continued performing their job. We were indebted to them for it and weathered the incoming fire to see them through.

Adrenaline masked our invisible wounds while scars became infected. I was emotionally inebriated with a misty view of the world outside the one in my hands. Clarity was a privileged commodity and unequivocally beyond reach. My heart burrowed into the cavern of my chest. I sank further into the unknown of my soul, fearful and intrigued at the depth of my unearthed being. However, the approaching holiday carried me out of the daily struggle to remain human.

Thanksgiving offered a renewal of spirit. Tensions within the platoon were forgiven in the desperation for sanity. Units serving across the city were given time to gather as a community where rank could be unpinned to receive the blessing of our homeland. The ability to celebrate in the midst of chaos brought meaning to our suffering. Our persistence in fighting for small victories saw beyond incidental grievances and revoked power from the demons taunting us.

Camp Ramadi's chow hall was decorated with evidence of joy and spirits rose in unison. Pumpkin pie, turkey, ham, mashed potatoes, gravy, stuffing and sparkling cider were available to everyone who marched through the doors. Plastic decorations and cheap red, white, and blue streamers brought color to a room that was painted with professionalism. Religious or not, we all gave thanks to God for preparing a table in a country where only Allah could be worshiped in a public venue without repercussion.

I tried to imagine how Marines in the Pacific sailing from island to island might have celebrated in the field. Resources were restricted due to constant travel and access to food and ammo depleted rapidly. In the Korean War, Marines faced similar logistical isolation but with frozen rations and frostbitten limbs. Vietnam was a humid nightmare where unit cohesion dissipated as Marines were replaced more frequently than chow. Camaraderie with their platoons was the most valuable commodity.

Marines in Afghanistan maneuvered through large open fields without cover, at high altitudes, and harsher weather conditions behind enemy lines where resources consisted of what they packed in. "Meal No. 18 - Turkey Breast with Gravy and Potatoes" was the golden ticket for celebrating Thanksgiving. Senior Marines across the battlefield pulled rank to acquire it. Sweat and blood paid for the land we called home. By all accounts, we faced an easier battle even at the peak of violence in the Anbar province. Though, our victory was beyond small. I was truly grateful to sit at the table with my brothers, paying reverence from the comfort of a chow hall. The pervasive negativity from three months of fighting fell silent over the course of a single meal.

It was the beginning of our fourth month in country with Christmas and the New Year approaching. Cooler weather helped recover a circadian rhythm and allowed our cynicism to recalibrate. Some nights I fell asleep without effort. Others were spent staring at the roof wondering when mortars would ventilate the insulated tin and shower liquefied steel in one cleansing pass. I loathed restless imagery for complicating something so routine. It was easy to become hopeless, to lose perspective from mental exhaustion and entertain my own demise.

I finally achieved enough fatigue to let my eyes fall. Sinking into the foam block that absorbed dreams like a sponge. My guard faded as my head depressed further into the pillow. Mental injuries subsided—muscle tension reducing. My heart was resting at a comfortable 65 bpm until the concussion from a large explosion ripped through the city unannounced. All sleep requests were denied.

The shockwave drew blood away from my extremities then my heart exploded with pins and needles through millions of veins. The adrenaline dump temporarily displaced my oxygen intake. The surge brought me back into a disgruntled wonder-land with a quick pat down for shrapnel wounds. A glance around the room affirmed a synchronized prayer. The insurgents weren't going to let the holiday remain victorious.

CAAT Red spent a few weeks rotating positions in the trucks to maintain familiarity with weapons and vehicle function. I switched positions with Schaefle to observe the city from a different perspective. Manning the .50 cal gave me a chance to focus on the uniqueness of Ramadi's structures other than roads. Patrolling the Souk put me on guard more than any of the surrounding areas. The business district quantified threats from elevated buildings and unreachable angles of fire from a mounted machine gun. Patrols through congested areas battled with low hanging wires against radio antennas.

We received a call to support an Army convoy hit by another catastrophic IED. A Bradley tank was actively burning in the street. We provided security at the ambush site to cover the Army's extract of the damaged vehicle and personnel. Combatants would attempt to utilize the downed vehicle and wounded men as a way of luring first responders into their fire. Our presence repelled the attack and gave them the chance to evacuate wounded soldiers.

We made our RTB and were clearing weapons when a flash of white hit the sky. I stood outside Alpha 4 hesitantly clearing my rifle and braced for concussion. QRF was activated and Marines from another CAAT team rushed to their vehicles. Within two minutes, they were staged at the clearing barrels readying weapons and equipment. I found relief in being surrounded by men who ran toward a fight. Their courage vindicated my decision to be there, affirming our loyalty to each other. Alpha section stood down and let an adjacent platoon handle business.

Details from the attack reached CAAT Red by evening.

Lima Company had been transporting Marines in a 7-ton truck that was disabled by a large IED. Marines dismounted from the vehicle to assess the damage and posted security when a secondary IED exploded underneath the vehicle. QRF was called out by the time a third IED detonated before their arrival. The QRF team held security while scanning the immediate area for incoming small-arms fire. Marines and Corpsmen dismounted to begin treating the wounded. Nine were badly injured and one was KIA.

Hostility carried over into the following day with tenacity. Task Force Wolverine was hit by a large IED. They reported enemy movement incoming from their position in the heavily populated Souk. CAAT Red was called out to provide a security screen near the downed vehicle. While en route, adjacent units radioed in enemy contact advising responding units to scan for MAMs throwing grenades and Molotov cocktails from rooftops. The possibility of being set on fire by a Molotov kept my eyes high. I sank low in the turret and scanned taller buildings with a M9 Beretta in hopes of deterring thrown objects. We reached the vicinity of the down vehicle and slow-rolled to interrupt enemy movement.

The Souk was a ghost town. Few people were walking the streets and businesses in the area were closing down to avoid further destruction. Streets were narrow and covered with low hanging electrical wires and ratty tarps for shade. The roads were caked with dirt and debris, garbage was collected into piles and randomly placed downtown. Sewage water pooled along the curbs and buildings were drilled with bullet holes. Ramadi's Souk in the Quatana district was a stereotypical front-page image of the war in Iraq.

I fixated on the possibility of fiery glass shattering across the truck, sending me over the side and exposing myself to incoming fire to extinguish the flames. We pulled down the road parallel to Bravo section and held security next to an alley. Civilians moved hastily on narrow back roads to avoid possible crossfire. Suddenly, three insurgents ran across the rooftop of a building at the end of the alley. Reis called out their position and I rotated the turret into place.

"Sanderson, hit it!" Clifton shouted.

I sighted in and slammed the butterfly on the .50 cal, sending four bursts of fire in the direction of their position. I waited for the dust to clear and assessed the target. Rounds shredded the slim alley without mercy. Armor piercing rounds cut walls on the lateral limits and tore through a bright pink building at the end of the road. Not a single person was visible on the street when the dust settled. The sound of gunfire sent everyone running.

CRACK! A single round flew over my head. I tucked back into the turret for cover. A sniper opened fire without any bearing of where it came from. A few seconds passed before Bravo's vehicles pulled up at the end of the alley directly in front of the pink building. I spun the turret to the rear of our vehicle and scanned the 5-6 story buildings. A loud *WHOOSH* disrupted my observation followed by an explosion 100 meters down my line of sight. Another RPG had missed its target. 7.62mm fire trailed behind the RPG from an unknown location. I waited for a break in gunfire and exchanged rounds in the direction of cyclic fire, hoping the armor piercing rounds would establish contact with barricaded insurgents.

Schaefle backed the truck in, closing distance on the where my rounds impacted. We staged directly between two of the largest buildings in the city beneath a shredded tarp. Just then the section leader called over the comms, "Alpha dismounts, post security and link up on foot."

All VCs and dismounts departed the trucks to sweep areas beyond the reach of vehicles. Drivers and gunners held security on the road. Visibility of the area was minimal from the buildings and even tighter from the turret. The .50 cal was useless with the restricted movement and tall infrastructure. I was left with a full-sized M-16A4 and side arm to cover enemy movement within proximity of our vehicle but I was sitting blind to anyone within five meters.

I sat with my eyes glued to balconies and windows remembering the threat of Molotov's and grenades. I reclined back to maintain vertical cover. I kept my M9 pistol at the ready, searching for any signs of movement. Our position was well within range to engage anyone that shot dirty glances.

"Schaefle, I'm completely blind to anyone on the ground near our truck. You gotta be my eyes on anyone approaching. Don't let anyone gain access to the doors." I called out in a panic to regain my bearing.

"It's clear right now. This is a tight spot though, My M-16 is useless from this seat and my K-Bar is the only thing accessible." He responded.

I started to collect myself while waiting for the rest of Alpha to finish their foot patrol of the area.

"Hey Sanderson, did you see that guy in the alley?" He chimed back in.

"The one with the donkey cart? He's the last one I saw before engaging,"

"No, after you fired."

"I didn't see anyone after the smoke cleared. I ducked to avoid the sniper fire. What guy?" I responded.

"Down the alley…" He was interrupted by radio chatter. "Alpha, be advised we are headed back to the trucks. Alleys look clear. We're moving to you from the east." A swift response from another vehicle confirmed their movement, "Solid copy."

Marines on foot took their place in the vehicles and we RTB'd to HP. We made it to MSR Michigan when the conversation opened to expel the bulging adrenaline rush.

"Right on Sanderson!" Gonzales called out.

"I don't know what the hell you were shooting at." Clifton shot back to discourage compliments. "Your rounds were all over the place."

"I'm free gunning the .50 Corporal. It was a small alley." I was struggling to hear the rest of their conversations over the diesel engine and ringing in my ears.

"Did you see that guy Sanderson?" Reis asked.

"What? No, what guy?!" I said feeling left out.

"Someone peeked around the corner at our vehicle right as you fired. A round went through the wall where he was standing and all I saw after the dust cleared was feet sticking out from the corner." He said.

"I saw him too," Schaefle added.

"I didn't see anyone." I responded.

I thought about the collateral destruction from the rounds fired and felt no emotional attachment. The rush of blood to my head was still draining. Sensory overload blocked anything from making sense. It wasn't until we departed the Souk that I realized my hearing was shot. Unleashing the .50 without ear protection went unnoticed until my adrenaline rush faded. The dampened sound added to the confusion. Only vague responses and chatter were audible over the vehicles.

The patrol lasted a total of four hours until all American personnel could be evacuated. We debriefed immediately upon return. I was required to give a statement on why I used deadly force. Dismounts gave their account of the foot patrol. They searched the immediate area for insurgent activity and hadn't found a body. It was possible that someone had recovered it before us judging by the blood trail.

The responsibility of fire was mine to own. Without seeing it, I was skeptical I had actually put someone in the ground. Reis approached me after Alpha's debrief and asked if I was alright. I sensed reservation in his voice and dug to uncover the truth of the matter. He looked as if he had seen something too damaging for me to know about and withheld details of an already menacing responsibility. I restrained my curiosity.

In the evening, a flare shot across the night sky traced minarets in a casual burst of rage followed by the growl of a thunderous breath sending me to cover. Gunfire ensued as tracers clawed at the black sheets above. I watched from a position of perceived safety as the spirit of Ramadi cried out in sorrow. Our rendezvous at the fire pit offered warmth in the easing cold of the winter months.

Passive sighs and blank stares into the fire converged on the space where we learned to appreciate simplicity. Nights without conversation exposed our deepest parts and allowed wounds to heal without the intrusion of conversation. The crackling of wood snapped like bones being reset to heal in alignment.

A few Marines returned from the phone center with updates on current events and a SITREP from buddies in adjacent units. One Marine received an email from of a friend in 2nd Battalion 7th Marines, operating in Fallujah. From our position in Ramadi, they were less than forty miles to the east. Three Marines we trained with in SOI were killed earlier that week. My original orders for 2/7 in SOI flashed through my mind. I was unsure if it was a blessing to be in Ramadi over Fallujah. I stared into the fire again searing the nagging doubt eroding my thoughts.

Our eyes glassed over in a sepia toned reflection from the flames. So many events were unfolding and denying the emotional appeals. I felt blank yet above victimhood. I was helpless to change anything but my state of my mind. Death was just another box to be checked. I grew indifferent to loss and felt nothing at the work of my hands. The brothers next to me were all I could think about. It was the only place I could be present at the expense of the outside world.

My heart was restless from the assaults of fear. Bravery was a strength achieved the harder my heart became. I immersed myself in the chaos around us and caught my reflection with contentment. Anger disguised itself as courage, enabling action beyond emotional impulse. The chance to return fire was a flag planted in the ground. I felt complete watching the dust settle though I was drifting toward the same ill intentions that put us here in the first place.

A quiet voice in the back of my mind spoke of the person I was becoming. I was more scared of myself than the enemy. Scared that I would give into the climate and lose sight of who I was beyond my time in service. Hope often came through a social jury, though I fought for a hope rooted in the eternal. I deferred my concerns to remain spiritually present.

Early in the morning, during another week of ECP, two drivers found themselves on the receiving end of our sights. Lance Corporal Penney and I shared visibility of the vehicle entry point at the Euphrates bridge when a vehicle approached the concertina wire along Route Nova. The lone driver ignored posted warning signs and attempted to proceed through the closed area. A single shot rang out as Penney called over the radio that he fired a warning shot to deter the driver. The driver retreated in frustration. Minutes later another driver made the same mistake. I fired into his engine block.

The guidelines in Ramadi were simple: No traffic comes in or out of the city between 2100hrs and 0700hrs. The purpose was to cut off weapons and explosives from being smuggled in the night. Both drivers attempted to pass ten minutes before the cutoff time. They were understandably enraged as a flood of traffic passed them on the same route they had just been denied.

"Post 3, can you see where the round impacted?" I called over my PRR.

"Ya, just above the headlight and he looks pissed!" Penney replied.

"Yeah, I bet your friend looks the same." I responded.

"Well that's two more IEDs we have to watch for now." He laughed as both drivers sat on the side of the road inspecting their vehicles.

"Should have read the sign before moving the road block." I said.

"General Order number eleven: allow no one to pass without proper authority." Penney closed.

It was nearing the end of another rotation of ECP. The draining hours in solitary observation gave me another chance to reflect on the place I was in and drink coffee in a state of peace.

The lack of electricity added to the décor of the building's dilapidated construction. A four-inch diameter PVC pipe supported by sandbags served as a urinal, running through a hole in the wall hanging over open ground on the outside. A small sand pit lined the floor of the urinal to catch anything that missed its mark. Squatting was done in plastic bags, which were taken to a burn bit. "Wag bags" were then soaked in JP-8 diesel fuel and lit on fire to keep waste from piling up. The stench augmented the air quality of the city.

The drastic shift from vigilance to primitive defecation tempered our pride. I sank back into existential reflection looking over the city. The weather shifted, bringing an additional victory in addition to charred coffee. Clouds poured over Ramadi with the same grey tone of city's foundation. Pensive indie music played in my head as the cool air swept past the bulletproof glass in front of me. The wind lightly rustled the grove of palms within my lateral limits, while a lone shepherd guided his flock under the cover of various date palms.

The road where the two drivers were left stranded bordered the palm grove and the Euphrates gracefully flowed along the north edge of the city. Behind me lay the Habbaniyah Canal filtering the weight of Ramadi's geographical misfortune. History was being written during our time in Ramadi. I wished I could know the full history of events that carved our place in Mesopotamia. I felt honored to experience Iraqi culture even through adulterated conditions. Standing watch felt timeless as the sky began to open. Thunder crashed unexpectedly, like an IED from the sky. A cry from the heavens brought the earth to life with a soothing petrichor. Simplicity owned the moment.

Rain was a peace offering to share in mourning. I was filled with gratitude to see Ramadi drenched in tears. I glanced down from my second story post to see the Iraqi shepherd sitting in the dirt. He brandished a cardboard sheet to cover himself from the rain. His goats and sheep grazed the courtyard of palms oblivious to the mysterious water falling from above. I was invisible in his presence while feeling connected through resilience. The coolness of the evening purged frustrations in the unscheduled assembly of a newfound humanity.

CHAPTER SEVEN
SMALL VICTORIES

MORNING BROKE ON DECEMBER 18th with a heavy-hearted platoon brief. Not long after waking up, Carson called the platoon together. His facial expression created walls around our spirits knowing they would be shattered once again. He regretfully proceeded to inform us of CAAT Blue's KIA.

"If you haven't already heard, Lance Corporal Tapia was killed."

Silence sutured the room with a familiar sting. Word of a fellow Marine's death became increasingly common, defeating an already weak barricade around our grief. His words would resonate for years and he bore that responsibility faithfully.

A cluster of mental resolutions flooded my mind and everyone else's. *Swanberg, Bedard, and now Tapia.* I couldn't process it. The sinking weight of my spirit threatened a complete loss of self in the absence of understanding. I took refuge through a rational mind as a means to continue our mission.

Some Marines remained silent as they tried to rationalize what Carson said. I searched the room for the subtle hope of contradicting evidence. Some were already in tears as others stoically denied emotional impulses. Carson continued, "CAAT Blue was conducting a raid on a compound. They posted security and dismounts searched the area. They cleared the roof of the compound and were headed back to the trucks when a sniper fired, killing him instantly. He was EVAC'd to Charlie-MED but there was nothing they could do."

Evil assaulted the platoon–extinguishing all hope. An unprecedented rage exploded within the room. Lance Corporal Curtis walked toward the back of the hooch compromising his composure. Penney lost resolve with a similar dejection as their outburst tore apart our hardened bearing. Their anger man-ifested before Carson could finish. A few Marines rushed to intercede. Penney gave in to their efforts but Curtis was beyond recovery. He grabbed his rifle, inserted a magazine, and chambered a round. With a loaded weapon and blinding tears, he screamed helplessly, "I will kill you!"

Carson interjected to maintain control, "We're all pissed Curtis. You don't think we understand? That the rest of us can't feel the same way? I want him dead just as bad the next person, but this ain't going to solve anything. Now, let me finish."

He went on, "Look, this guy knows what he's doing. Word is that we're dealing with a Syrian sniper who is helping the insurgency here in Ramadi. He's obviously trained and was able to make a headshot on a moving target. I don't know how far, but regardless, it's a precision kill."

He paused to swallow his emotions, "Gunners, keep your heads down in the turrets. Don't stand up if there is no reason to. Dismounts, move quick and maintain cover. Stay off the streets. This isn't anything new to you guys. You know what's out there. This is a serious threat. It doesn't take ten Marines getting killed to feel pain, it takes one. And don't bottle this up, talk to each other about it."

I focused my attention on the platoon to gauge an emotional response like a mirroring child. Some were still fighting tears as others were frozen in shock. Others excused themselves to have a private moment. Those who remained stoic began to worry me. I tried to console Penney and Curtis, but nothing I said would inhibit the momentum of this loss.

Penney forced words through a broken spirit, "Tapia planned on calling his family to wish his oldest a happy birthday today."

I sympathized more for them than for myself. The depravity of hope was a festering wound with seemingly endless limitations. Our vulnerability was masked by the massive strength we embodied, but at our core we were fighting to remain human.

Everything we endured substantiated my decision to be there. The less human we became the more we began to understand our nature. Ramadi was taking more than I was prepared to give. Sacrifice demanded more than flesh and blood. Mind and spirit were tarnished at the expense of the future. The distance from home grew further every day we survived.

A funeral service was held a few days later and a somber glow fell upon the entire company of Marines present. The final roll call, twenty-one-gun salute, and playing of Taps aligned our silence yet again. Some of us left the room directly after formation to make peace with each other. Trivial conflicts within the platoon were forgiven through emotional fatigue.

I approached Penney who was failing to contain himself. He kept his head down and opened his arms in submission to the fight. He broke down in my arms, transferring tension onto myself. Everything I had been holding back in order to save face in front of others was lost and I buried my face in his shoulder in return.

"I love you brother," he spoke with weak posture.

I broke character and responded the same. The depth of loss reached beyond the exchange of words. Aggressive posturing was a façade in the presence of suffering. We both understood how precious life was. Our next patrol could be our last and soon enough we would find ourselves revering the battlefield cross again. Grief broke feelings of insecurity and gave meaning to our suffering.

Christmas beamed as another milestone of restoration. Family members sent gifts and decorations; the pieces of home we needed to change the atmosphere and state of our minds. Christmas trees emerged from care packages at the beginning of December and colorful bulbs hung along the dark seams of our tin roof–streaming life over a pale and fleeting backdrop.

Christmas promised a soft close to the year with a month of sustained victory while the enemy was eager to steal our seasonal joy. I called home hoping for relief but something was still missing. I managed to get a few hours of rest after a short patrol on Christmas Eve, but I was failing to recover. The separation from home and the nature of our work expanded inside my skin, tearing at weak points. I was homesick in the pain of growth, at war with God.

My first conscious hour of Christmas came at 0200hrs during firewatch. Chilly weather filled my spirit with a new love of life. The rest of the platoon fell asleep under the dim safety of colorful incandescence. Christmas trees draped with tinsel and ornaments conquered the back of the hooch emanating through the cracks of weak construction. Christmas revived my broken spirit and ushered in a gleaming hope for the New Year.

I was relieved of watch at 0300hrs. Unable to fall back asleep, I sat near a small tree and drifted off into my headphones. Acoustic guitar strummed on repeat and directed my mind with grace. Suddenly, I was back at home driving with Kristi viewing the Christmas lights in the quiet foothills, listening to carols of songs that defined us in high school. I wanted to hear her voice. I needed her near me to restore what I lost. I saw her in the beautiful state she once existed, from my place of darkness where she would neither recognize nor desire me.

I opened my eyes to the dim room and tiny tree. My own mind was powerful to sabotage. The joy of Christmas removed me from a world of chaos to the ethereal space of the past. Music poetically narrated a life of wonder with visions of a future after the war. The place I had come to fear the most wasn't Iraq, but the vision of a life in vain.

I was asleep and up again, ready to go through packages before the platoon awoke. After spending the morning with Josh, I went to find Mike a few hooches over to wish him Merry Christmas. The short walk through the rain covered more than a multitude of sins. Small victories crushed the giants of anxious fear. We shared Christmas morning together over 7,000 miles away from home. This was family, forged through bloodshed.

New equipment was fielded as a late Christmas gift for American forces in Ramadi. A Counter-Battery Radar system designed to detect enemy fire was set up to cover the proximity of FOBs on Ramadi's west end. Indirect fire and rockets could be identified before they could impact their targets. The system would calculate the trajectory and location of fire down to a grid coordinate enabling return fire on the exact position. A loud alarm would sound to alert personnel of the incoming fire and gave us a chance to take cover before impact.

Counter-battery radar underwent its initial testing phase two days after Christmas–none of the Marines in CAAT Red were informed. The alarm wailed across the Hurricane Point through a loudspeaker: "INCOMING! INCOMING!" Marines across HP stopped dead in their tracks. A few seconds passed before common sense kicked in and sent them running for cover in the closest shelter. No rounds impacted, causing further confusion and distrust for the new warning system.

The following day, junior Marines from CAAT Red were voluntold for various working parties while on secondary QRF rotation. Harkins and Schaefle won the working party lottery from Alpha section and were sent to the Battalion Aid Station (BAS) for building maintenance. While standing outside of the BAS, the audible alarm sounded off again: "INCOMING! INCOMING!" Thinking it might have been another test but not willing to wager their lives, they bolted inside for cover forgetting their rifles outside. A few other Marines narrowly managed through the door before a crippling blast uprooted the dust beneath their feet. The blast was in close proximity.

A Captain stumbled through the door, hands cupping his neck from shrapnel. Marines yelled for a Corpsman and gained a quick response. After receiving the "all clear" Harkins and Schaefle proceeded outside to retrieve their rifles only to find a crater in place they were standing. Their rifles were maimed by the blast, rendering one ineffective.

Harkins and Schaefle rushed back to the hooch with their broken gear for accountability. They stumbled into the hooch yelling, "You're not going to believe this!" The first operational day of the radar system substantiated its effectiveness and gave everyone at Hurricane Point a greater sense of security. Harkins was issued a new rifle and Schaefle discovered damage to his video camera. He would have to cash in on the warranty. My impatience in hearing from God was shattered in the unbelievable miracle, brushed off as a lucky coincidence by others.

New Year's sent Weapons Company into a downward traverse on a mountain of bones. Ramadi cut wounds so deep that a majority of us would remain lost in the lacerated crevices of the city. New Year's resolutions were the past time of a protected society and a naïve nuance for Marines along the Euphrates. Our efforts in Iraq spilled onto the pages of history, where clotting blood held the binding of a living document.

Patrols habitually enticed us into contact with the enemy. Stagnation was the devil's playground where minds wandered into the snare of complacency. Patrols longer than eight hours demanded a mental endurance that was depleted by the minute. Vigilance could only be sustained for so long. Presence patrols were a mechanical endeavor of baiting insurgents until bombs went off and rounds started to fly. By then, our wonderful plans were unserviceable as our equilibriums recalibrated from the hard reset of concussion.

CAAT Red received a few "hit and run" missions conducting raids and gaining trust with the locals. I maintained security at the first house we came to waiting for the call to secure the next compound. While staged, a little girl peeked out from the front gate of an adjacent house. Assessing the area and determining it was safe, she stepped out to serve a better glance.

I flagged her attention and motioned to come closer to the truck. With a fistful of candy loaded in my palm, I cracked my door instead of throwing it from the turret. I nudged the door open as she glanced at me in hopeful suspense. Her face lit up with a smile when saw the candy stretched out in her direction. She snagged all she could and ran off without a word.

I contemplated the kind of impressions we were leaving. I tried to appreciate the seemingly insignificant moments to disrupt a corrupted view of humanity. It was a brief escape from reality to interact with kids. It gave us the opportunity to clarify our image to the future of Ramadi. We were always ready for a fight, but most of us hardly felt above interaction with the locals.

In a city dictated by strict religious and cultural norms children weren't expected to act like their adult counterparts until their teenage years. Their interaction with us was loosely restrained: making eye contact, communicating, and hustling for chocolate and soccer balls. Iraqi kids substantiated the belief that regardless of country or region, children have the same spirit of wonder. Their innocent wasn't lost to a destructive society, but neither was it immune to the harmful effects.

The mixed group of insurgents exposed a weakness within Al-Qaeda. Foreign fighters killed anyone for their agenda and on occasion, cracked off rounds and RPGs with the locals downrange. Insurgents risked collateral damage at the expense of their own countrymen. Killing Marines was their priority, but they were not above coercing women to participate in violence by driving vehicle-borne improvised explosive devices (VBIEDs) and concealing weapons. Iraqi nationals fighting the insurgency avoided killing their own. The conflict of interest began to erode the strength of their numbers.

CHAPTER EIGHT
FALSE HOPE

JARRING OUT OF ANOTHER NIGHT'S sleep, I was welcomed back into the world with the uncanny feeling that my darkest dreams were unraveling. January saw a vibrant morning light and gentle sewage breeze that debased Ramadi's vintage sphere. Evening skies hummed on the horizon in an extravagant display of colors this grey city desperately craved. Alternatively, inflamed morning skies signaled hostile conditions and nursed a bitterness that hardened my heart even with collaborating swirls of red and yellow.

The sailor's tale of evening red was more than vindicated. Chaos left an appreciation for anything that filled life with remaining color. The camaraderie building in our trials made Iraq an unexpected sanctuary. The duration of time spent in Iraq was bearable for every Marine and became preferable to training in 29 Palms. Everything we did bore a consequence, giving meaning to our work; from no-knock raids to candy donations. I was fulfilled in how far we had come and was learning to enjoy who I was becoming on the descent of our deployment.

The fifth day into the new-year proceeded in cyclic fashion. The locals were beginning to align with our purpose in Ramadi. Iraqis independently came forward to serve with a personal investment for their freedom. Next to Hurricane Point was an abandoned glass factory where a recruiting station had been established to bring in volunteers for a new police force in the city. Ramadi was collapsing under the corruption of their former police and military force.

The American presence was minimized to allow Iraqis to own the progress of healing the Anbar Province. A senior ranking Army officer and Marine K-9 handler oversaw the recruitment efforts to help screen qualified men. Locals arrived with a restored hope of defining their future. US involvement in the recruiting process provided oversight to ensure integrity and prevent corruption from resurfacing.

September felt like yesterday as we came closer to our departure. We were finally seeing the fruit of our labor. Alpha section staged at the clearing barrels awaiting to depart for a routine mission. Untold delays outlived our patience and idle time carried our minds down the canals of complacency. Uninformed stand-by depleted the aggression that naturally fueled my vigilance. The sunflower seeds and Rip-it energy drink I pocketed on the way to the trucks were the only joy I looked forward to on patrol. Snacks were a glorious remedy for the disgruntled. But I was going to need more than Rip-its to keep my heart in step.

A concussive wave stripped away my complacency in the driver seat of Alpha 4. *That was close!* The dirt on the trucks leapt a foot into the air and everyone took cover under a cloud of suspended dust. A large bomb detonated just beyond the walls of HP near the glass factory. Marines on the ground hastily recovered to their feet, assuming their positions in the trucks. QRF was gearing up in response. Alpha was still waiting for clearance to push when the decision was made to send us out as QRF instead, placing the activated team on standby.

My excitement grew larger than the guilt of desensitization. Breaking from routine to hostile activity meant history was being written. None of us knew what was unfolding, yet we salivated with lust for the fight. I fell into the hypnosis of curiosity that could only exist on my side of the wall; a victim to puerile emotions and neglected prayers. All of us would be permanently affected.

"Red Alpha you are clear to push!" A voice screeched through the radio. We pushed through the gate like a greyhound chasing a lure. It took seconds to reach the glass factory but our chances of fighting were reduced to CASEVAC procedures. A quick assessment of the scene revealed that a suicide bomber infiltrated the crowd of Iraqi volunteers decimating over forty lives. The displacement of body parts across the blast site meant the explosive vest was packed with ball bearings. The added shrapnel achieved an even more heinous effect.

Bravo Section provided additional support. High-backs pulled inside our security cordon to evacuate the wounded Marine and Army Colonel. CAAT Red began triaging the wounded, loading bodies into the high-back for transport to Charlie-MED hospital. Marines and first responders cringed through internal repulsions to reach them. The screams of the mangled were given priority as the silence of others were granted peace. The mass of casualties activated support crews from Camp Ramadi and calls went out to clear more space at Charlie-MED. All available vehicles were rushed to escort the living. The short distance from the glass factory to the medical facility at Camp Ramadi was still too far for some.

The haze of fear and death saturated the air with malevolence. When the only sounds left were first responders, it was clear that our intimacy with death was consummated. All boots on the ground were handed black bags and ruefully disposed of the deceased, piece by piece. The smell of charred bodies and the echo of lost voices castigated innocence beyond recovery. Time held priority due to further threats of violence, causing fingers and toes to be casually buried in the sand. For some, the physical interaction with torn flesh activated the vile in their stomachs.

When all casualties were treated, CAAT Red filed back to Hurricane Point and was given a substantive pause before debriefing. The platoon ushered in with no regard for time. No tears, consolation, not a single word carelessly spoken. The room loomed under the weight of speechless sorrow and the smell of burnt flesh. The Platoon Commander entered the hooch disrupting the air with his presence. We sat in the wake of a new stillness–each man feeling alone in the crowded room. After a brief moment, Lt. Awtry addressed our shattered innocence.

"For those who didn't witness the aftermath, count your blessings. This is one of the worst things you'll see here..."

I trailed off into the embrace of shared misery, feeling betrayed by my desire for malice. The thread dividing imagined violence from actual violence was indistinguishable. The massive loss of life challenged the value of our business. The depth of darkness surrounding us wouldn't change our dedication to the mission at hand. It was clear who the enemy was and the Iraqis did too.

The internet and phone center reopened two days after the attack. Josh and I were catching up on the way to contact family at home. The somber nature of his usually cheerful demeanor caught my attention.

"Hey man, are you good? You're not usually this quiet?" I asked.

"Ya, I'm just tired." Josh responded vaguely.

I resumed conversation, "Alright. So hopefully the Internet center is empty cause I want to send some pictures out from this thumb drive again without getting caught. I have some cool shots I took from patrol."

"Cool," he said with a sigh, merely acknowledging that I had spoken.

"You should see them, I have a few of..." I paused momentarily, "Are you sure you're alright?" His disregard registered louder than my voice. "Do you need a cigarette or something because you look antsy?" I leaned into his mind.

"No, I don't need a cigarette, I want to get out of here!" He slammed back disturbed. The force of his outburst was unusual.

"I get that man, we're almost out of here. More than halfway through this deployment." I welcomed his frustration but missed the trigger.

He continued, "I'm sick of this desert, and not sleeping, getting blown up, and standing watch every night."

His frustration gained momentum beyond the surface of deployment fatigue.

"Let's just call home and see what's going on back in the States. It helps to hear from someone other than another Marine." I tried shifting his focus but he was beyond recovery.

"Why, so they can ask me how things are going? So, they can ask me about the suicide bomber they probably heard about on the news and if I'm OK? Well I'm not OK! I don't want to talk about what's happening here. I don't want to explain to my mother that I watched my VC pick someone's head off the ground and dump it in a trash bag. Or, that the Soldier they loaded in the bed of my high-back died while the Corpsman tried to save him. Did you know, he was screaming when they loaded him in the back, but was silent when we arrived at Charlie-MED. I don't want to be here anymore."

He lost all bearing and broke down trying to piece together the frail remains of his humanity. My self-pity diminished in his struggle to stay grounded. I had the compulsion to interrupt the tragedy of seeing him fight to regain himself. The thread of sanity that held me together extended to him in a silent embrace.

It hurt to watch him suffer outwardly with the same pain I kept buried inside myself. I felt weaker in his strength to burst openly. He reached his limit like I had with Kristi before deploying. Both of us were desperate for relief. Being present was all the grace I could offer.

Fluorescent bulbs hung carelessly across local store entrances beneath torn awnings and ruptured buildings. Streetlights interrogated cinder block walls and metal gates that lined the fatal funnels of residential back roads. The blend of artificial radiance improved observation but whited out NVGs in direct line of sight. Iraqis closed their shops and sat along sidewalks smoking hookahs. Groups of people could be found watching TV or socially congregating in public areas and pay little to no attention to our presence. A courteous wave from time to time treated us like casual traffic.

Drivers and gunners paid the most attention to their surroundings, while dismounts dozed off in the vehicles on long patrols. The weight of my Kevlar helmet created another splitting headache, further contributing to disgruntled banter. I swung my PVS-14s to my right eye to offset visual corrosion from prolonged light exposure in one eye. The lime tint of the optic became rich forest green to a fresh pupil.

Positions in the truck shifted slightly before the end of our deployment as Sergeant Carson reached his end of active service (EAS) and was shipping out early. Staff Sergeant (SSGT) Yellope filled in as his replacement. Clifton was transferred to Bravo Section and Corporal Quinn became the VC of Alpha 4. The entire truck was relieved from the burden of his emotionally unstable attitude. Our remaining patrols allowed us to focus on the mission rather than narcissistic backlash. We topped off on coffee and pushed out for a long evening in the city.

Running into the early hours of the morning was stealing our hope for a swift RTB. I slowly trailed Alpha 3 around each corner in the ninth hour of our patrol before returning my attention to an empty coffee mug. My back curled under the pressure of ballistic plates where the edge of the rear plate sculpted a knot on my spine. I was safe from small-arms but was being maimed by safety gear. My mind gave way to exhaustion again. *I wonder what I would be doing in college right now...*

Minor existential crisis' during presence patrols surfaced with more pressure than the enemy. I battled against outward expressions of discomfort but lost all bearing.

"What are we doing? Is there even a point to being out here any longer? I'm starting to lose my mind!" I announced rhetorically.

"Starting?!" Doc chuckled behind me.

"Really, Sanderson. Should I call over the comms and let them know we should push back because you need a break?" Quinn interjected.

He was a quiet person, but had no problem letting me know I was still a junior Marine. My mind continued to wander in a primitive state of being. Before I could scold myself on the life decisions that led me to this point, we began pushing toward HP. *It's about time*, I thought self-righteously while mustering all the energy I had to finish off the patrol.

We rounded a corner, bounding north then west across the city, maintaining a tactical route to avoid the complacent posture of the most direct route. "Alpha 4's up," Quinn called over the radio to advise the other vehicles that we were on the same road. I accelerated up to our typical dispersion from the Alpha 3 when the concussion hit my truck. A bright flash was followed by smoke to the starboard side of Alpha 3.

Quinn called over the radio, "Alpha 3 is hit." We were closing the distance behind them. Adrenaline flushed all fatigue. Alpha 3 was trailing off to the right and dropping speed. *Oh no*, I thought to myself. Something went terribly wrong. Their vehicle drifted off the road and came to rest with no radio response or movement.

"Alpha 4 posting rear security." Quinn relayed over comms. "Schaefle, keep that gun on our six."

"Roger Corporal." Schaefle confirmed.

Alpha 1 and 2 pulled back to cover other angles of attack on our downed vehicle. Dismounts from Alpha 2 poured out to assess the damage. The VC's lifeless body fell out from the passenger side when the first Marine opened his door. He nearly hit the ground before another Marine caught him and sat him back up. He was unconscious as were the rest of the Marines in Alpha 3. A few minutes passed before they restored consciousness.

One Marine's flak caught a piece of shrapnel but he was coherent. The gunner shook off the blast after stumbling around a bit. The ballistic glass of the VC's window was severely damaged by a large piece of shrapnel but effectively prevented another KIA. Runflats maintained the vehicle's ability to drive. They were lucky to escape with only a concussion. I waited for the thumbs up that everyone was good and that the vehicle was capable of driving back to HP on its own.

We made it back to Hurricane Point fully conscious from the cortisol spike. The Marines of Alpha 3 brushed off the attack as an inconvenience. Our nerves were beyond shot. Alpha's debrief included intel that Kilo Company also hit an IED during the night and EOD uncovered roughly ten more on a clearance run. A total of twelve IED's had been taken off the street in a single night. Fear diminished from the lack of basic human needs being met. Alpha was heavily sleep deprived. Long periods without food also set our minds to less important aspects of a security patrol. Sleep was the only way to reset adrenal fatigue.

I'd become familiar with truck watch since my arrival in the fleet. The infant hours on the morning of February 18th acquainted me with the quiet time I needed to unfold my thoughts. We were approaching the final month of our deployment. CAAT Red resumed QRF per rotation. The night was cloaked with the chill of a desert winter preserving the tragedies of history. I paced between staged vehicles hoping to repair a curious faith.

The transition from high school to the infantry wandered in my mind. Who I was becoming challenged my perception of a "good" Christian. I drifted further in thought about what I could have done with my life to be living the right way. The thought about where I would have been if I hadn't joined the Marine Corps nagged in opposition to my current circumstance.

Would I be sitting in a classroom right now?

Would I have been better off staying with Kristi?

Would I have become more faithful in a different pursuit?

108

I wrestled with preconceived ideas about faith and my role as a Marine. A furious guilt was brooding in my mind. I needed God's reassurance through the instability of doubt. I was navigating the balance of faith and violence and learning to follow through with my commitment to serve. My thoughts trailed farther from a loose imagination.

I took a deep breath to trust the stillness of God's presence. The Euphrates was just on the other side of the brush and concertina wire that lined the river's shore. Ramadi held so much history I knew nothing about. My soul slipped into the embrace of the Spirit in an attempt to dress internal wounds. I lowered my head in prayerful reflection before the darkness brought me back into the fight.

A deep *WHOOSH* trailed by an updraft of wind slammed into my chest. The explosion disrupted the night's reverence. I regained my bearing behind cover. No other Marines were within eyesight. The bomb's low echo placed its signature a few miles out near the center of the city. In less than a minute the COC door burst open.

A frantic Marine rushed out in my direction screaming "QRF!" I bolted inside the hooch ahead of him relaying the call, "QRF, get on the trucks!" Alpha Section grabbed their rifles and scrambled to the trucks. Vehicles were fired up and heading to the staging area near the front gate while a SITREP was being transmitted. Platoon leaders came together for a hasty mission brief before pushing into the streets. Staff Sergeant Yellope relayed the details to Section Leaders and VCs:

"CAAT Black Yankee Section was ambushed and is currently taking small-arms and RPG fire. They've sustained casualties and requested immediate QRF at the intersection of Sunset and 5th Street. Zulu section is en route to their location. Provide fire support and evacuation as directed. Load up on your trucks and brief your men on the move. Stay tight and move quickly, we're pushing." We were cleared before we could close the Humvee doors.

Alpha Section roared down the MSR as fast as our diesel engines would allow. With RPMs redlining at 60mph, I slid around corners just shy of flipping the vehicle. My heart climbed into my throat as I maintained dispersion with the vehicle in front of me. We avoided the most direct route in fear of ambush and secondary attacks. We diverted from the MSR taking alternate routes to Yankee's location, even though CASEVAC was imminent. We would be no good to anyone if we were dead.

We made it to Sunset Road and secured the area. Small-arms fire ceased and Zulu Section handled primary evacuations. Two Marines were immediately rushed to Charlie-MED. The remaining casualties were loaded into the open seats of our convoy. We pressed on to Charlie-MED and dropped them off before returning to the scene for further support. When all Marines were accounted for, we exfiltrated (EXFIL'd) back to HP. We held at the clearing barrels waiting for the stand down order. Both sections from CAAT Black staged next to us consumed with a grief we'd soon share.

CAAT Black Yankee maneuvered south on Sunset and turned down 5th street when their convoy was hit. Yankee 4 suffered a direct RPG impact from behind as they rounded the corner. Radio contact was disrupted and placed the lead vehicle in a panic to figure out what was happening. Corporal Matthew Conley, the Vehicle Commander, dismounted to re-establish communication with his section in person. Insurgents ambushed the convoy by detonating an IED directly beneath Yankee 1's position. Once Conley dismounted, the IED was triggered.

The blast lifted the Humvee off the ground, violently jarring

Marines inside. The blast shield behind the rear starboard passenger seat blew off its hinge, knocking Lieutenant Almar Fitzgerald, the Platoon Commander, unconscious. A civilian reporter in the adjacent seat was ejected from the vehicle. The gunner and driver were conscious but disoriented. Small-arms fire erupted after the blast. Marines repelled fire with mounted machine guns and M-4s.

Corporal Conley was killed instantly while Lieutenant Fitzgerald remained in critical condition. Cory and a few other Marines were wounded but survived. I found Mike while staged at the clearing barrels. The attack had removed the person I knew so well.

"Hey, are you good?" I asked with caution.

"Ya, I'm good." He returned with nothing more.

His response affirmed my concern with the only words he was capable of speaking. I nodded knowing it was anything but "good." Alpha Section departed Hurricane Point to retrieve the Marines at Charlie-MED. Cory was among the few who were treated at the BAS. His arm was wrapped in a sling as we pulled into the lot.

"How bad is it?" I inquired, not understanding his medicated state.

"Dude, it hurts so bad. They pulled some gnarly pieces of metal out of my shoulder." He sounded fatigued. He lost some mobility in his shoulder and was issued a sidearm in place of his rifle.

"Well, you're still here." I closed, giving him a chance to doze off from the medication kicking in. He asked how the other Marines were doing, showing more concern for the Marines in his section than himself.

I checked in with Mike the following day. He was struggling to recover. OG was also with Zulu Section and part of the primary CASEVAC for Yankee. OG's usual comedic persona shifted in reverence to the atmosphere. CAAT Black suffered the biggest loss in Weapons Company from a single attack. Lieutenant Fitzgerald was flown out of Iraq to be treated at a facility in Germany. After only a few short days, Weapons Company received word of his passing before he could make it to the States.

A memorial for Conley and Fitzgerald was held at Hurricane Point. All available Marines in Weapons Company were present. Bitterness stained my heart. I stood in formation listening to their eulogies, wondering if it was their own words. We had come to know each other's stories so well that another Marine could have given an honest account of their lives without reference.

The air was heavy as if smoke was all that filled our lungs. An inextinguishable rage burned to keep some from falling apart. The ceremony concluded in a final roll call of their names, followed by a twenty-one-gun salute and the playing of Taps. Marines lined up to offer their respect for the fallen Marines. Each Marine moved lethargically past their battlefield crosses, placing a hand on each Kevlar. Faces buried in the ground rose only to gaze upon the picture at the foot of each memorial.

Corporal Conley and Lieutenant Fitzgerald's memorial was one of the few times I'd been able to bring out tears. No amount of death ever became easier. No amount of violence could settle our pain. Marines understood what it meant to sacrifice. Blood outpoured tears in Ramadi. The rage fueling our violent endeavors tapered off with our return to the States approaching. Our efforts in Ramadi came to an end, but would not be the final chapter of our story there.

CHAPTER NINE
RESOLUTION

THE DAWN OF OUR TIME in Ramadi closed in the bloom of Spring. Downtime filled our remaining days, decompressing aggressive tendencies. Blood stained the streets like discarded business cards. Marines joked about what their first act would be in the States. Alcohol and sexual references exceeded the limits of social order. The duration of our flight passed without a sigh of inconvenience.

We stepped off the bus in 29 Palms into the crowd of anxious families. My lungs expanded, pulling in the cold air like helium. It was lighter than the toxic fumes we'd inhaled for seven months. I embraced my family for a while prior to exchanging words. Before driving home, I recognized a woman in the crowd. Swanberg's mother followed through to welcome our return. She welcomed all of CAAT Red as if searching for a resolution we all received. I was filled with guilt for the joy we shared.

I made it back to LA, attempting to entertain social conversations with family. Current events and local news updates were soothing chatter behind California traffic and formidable infrastructure. I grabbed the keys to my truck and set out to process confused emotions. I didn't know where I was going, but I needed to be alone. I parked in a familiar lot with a view of places I knew well. People moved about in utter complacency.

I was a ghost with the knowledge of a thousand lifetimes. The demons of Ramadi sat quietly beside me revering their destruction. I was vacant, breathing out of compulsion. The freedom in that moment was lonelier than any night along the Euphrates. Ramadi became a place of refuge from the entitlement of freedom. Materialism invaded my sacred space through advertisements and billboards. The presence of competitive carelessness was nauseating. My heart was hardened.

My eyes whipped across the room in a panicked search for my M-16. Spring mattresses and non-government issued linen couldn't prevent the instinctive reach through open air. The surfboard in the corner curbed a breaching anxiety. The presentation of colors in contrast to flat earth tones oriented my mind. The sun rose again as if time had not passed and a dream was all that transpired. Light passed between the blinds, refracting off particles of dust drifting back onto the sheets.

Post-deployment leave gave me time to explore the city I left behind. I searched for old friends to escape intrusive thoughts. I felt disconnected, with a deep need for human interaction. All that mattered to them was that I had returned, but I wasn't home. I swerved around potholes and sewer lids, aggressively changing lanes to get around trash piles on the side of the road. The restricting flow of traffic and hostility of drivers violated my personal space. SUVs and trucks were a clear indicator that VBIEDs were not a threat.

I was living in two worlds distinguishing between roles. A few surf sessions and the chance to adjust eventually settled my spirit. Things were not the same but I felt an overwhelming sense of accomplishment. I began to miss the Marines I'd spent every day with for seven months. The fact we had experiences no one else shared meant our conversations served as a social resupply. I headed back to base after post-deployment leave reset and ready to receive the next generation of Marines in Weapons Company.

We were back in 29 Palms in no time. It would be a year before our next deployment instead of the typical seven-month rotation and we were thankful for the extension. We transitioned from unconventional to conventional warfare training and dispersed Weapons Company back into MOS platoons. CAAT Teams were broken down into Machine-gun, Mortar, and Anti-Armor/demolition platoons.

A Company announcement was made that PFC Lira had been located, and brought back to 29 Palms. A Marine Corps personnel recovery team arrested him during the course of our deployment. We were likely to see him in passing. He was considered a deserter under the Uniform Code of Military Justice (UCMJ) for avoiding our deployment and would be pending Court Martial in the next few months.

He was placed two rooms down from mine and next-door to OG. A Marine escort dictated his every move. We rarely saw him outside of handcuffs. I had so many questions. *Why did you do it? Why didn't you come back?* No more than an occasional greeting was all I could bring forward. I had sympathy for him and risked facing consequences for small talk.

I didn't believe he was afraid of deploying. I understood his need to be home with his wife for the birth of their daughter. I even shared in his frustration at having his leave request denied. The few Marines who noticed him recognized what was left of his dignity. His eyes were filled with pain and frustration. I couldn't tell if he would quietly accept the conviction or if he planned to flee as he did before. I buried my empathy and placed my focus on an upcoming field OP.

I prepped my ruck for a field exercise with Kilo Company and verified my gear list with Mike. The transition to 81's platoon landed us in the same section. Our conversation was rudely interrupted by the smell of Lake Bandini.

"Was that you?" I accused him.

"Hell no!" He laughed then gagged. "It's probably Bandini coming through the plumbing again. It was like that yesterday too. I think it's getting worse."

My frustration peaked, "I don't remember it smelling this bad. They must have messed something up in these new barracks, it's probably someone's toilet. Smells like Ramadi came back with us."

Mike laughed at my helpless disapproval. "Well, at least you're going to the field for a few days so they should have it fixed when you get back. I had a hard time sleeping last night because of it."

"I hope so." I said as I finished packing and stepped out for fresh air.

I loaded my gear in the 7-ton and convoyed out to the range for the live fire exercise. Coordinating fire on mortar ranges was the best part of training in 29 Palms. It was less physically demanding with a heavy emphasis on mental coordination.

When rounds hit their target, the credit was in large part due to the accuracy of the Forward Observer.

There were only four FO's in the Battalion. All four of us worked together on Mortar ranges during 81mm shoots. I was attached to Kilo Company and got to know a handful of Marines including their seniors. They told stories of Ramadi from their angle and their deployments before in western Iraq. Some served with Corporal Jason Dunham, the first Medal of Honor recipient in the Iraq War, and talked about the kind of Marine he was. I was honored to serve with a unit writing the pages of history.

As a Forward Observer, I had a valuable skill that would be passed on to the next generation of Marines. Senior Marines were nearing the end of their contracts and entrusting the legacy of vital skills onto us junior Marines. They never ceased to impose their authority but respected us after Ramadi. We were disciplined and ready. Ramadi was our testament to them and to the incoming Marines that we were worthy of a good fight.

My field Op with Kilo ended and I packed my gear for transport back to Mainside. I looked forward to a shower and Santana's Mexican food from out in town. I turned my phone on to find a message from one of the senior FO's.

"Call me when you ENDEX [End of Exercise]." I called to let him know I was heading back to the barracks.

"Corporal Burge, this is Sanderson. I received your message, I'm in route back to Mainside."

"Alright, sounds good. The platoon is cleaning weapons at the Armory so you can link up with them there when you turn in your weapon. How was the shoot?" He replied.

"They're impressed with the accuracy of our CFF and were able to get a good amount of training in. Just trying to maintain a good reputation for Weapons Company." I said, pleased with the overall accomplishment. Experience came with a level of autonomy.

"Glad to hear, must have had a good instructor." He interjected to compliment himself. "Also, we figured out what that awful smell in the barracks was so you don't have to worry about it when you get back."

"Thank God!" I said relieved while remembering the stench. "Was it a busted sewage line or something?"

He chuckled with a desensitized breath, "No not a sewage line. Lira hung himself in his room a few days ago..."

A chill brought the hairs on my neck to life with an unsuspecting blow. I understood clearly what he said but failed to register the reality of what happened.

"What?" I replied.

"Yeah, he hung himself in his room. Apparently, the escort that was supposed to check on him daily dropped off for a few days. They didn't find his body till now so, he's been decomposing. Corporal Muniz found him in the closet and had to cut him down. Blood was pooled on the floor so they threw that salt stuff in there to kill the odor."

"Are you serious?" I reconfirmed.

"Yeah, crazy. Anyway, I'll see you when you're back." He replied.

"Alright." I responded, dropping the formalities of rank.

The words *hung himself* pierced over and over with a diminishing effect. My heart was silenced again. My suffocating emotions were fighting for air.

Why didn't you just come back?

Images of him retreating somberly into his room were the last I had. I stepped out of my head to avoid the pressing desire for answers. I arrived back at Mainside with no memory of the drive.

*   *   *

Our first drop of new Marines had arrived. We were finally seniors. Some carried a chip on their shoulder, treating the new Marines with the same contempt they were treated with. It served as an outlet for the persecution we faced since our arrival to the fleet. Having junior Marines around to task out for working parties or vehicle inspections was a glorious benefit. It alleviated our workload and vindicated our status at the same time. They were ours to delegate responsibility to and prepare for our next fight.

They were met with intimidation and fear tactics to see who was mentally strong and who caved under pressure. OG capitalized on their disoriented arrival. He stood in his doorway with a questionable grin while I found the key slot to my door. I dropped my pack and asked what he was smiling about. His face lit up as he shot a comment under his breath,

"I'm gonna tell them about the haunted room next door."

"What are you talking about?" I replied.

He carried on, "It's haunted, you know, with Lira's ghost. I'm gonna tell all the Boots that's where he died. You can hear his screams at night!"

"Wow dude, don't you think it's too soon?"

"No, it's perfect, it's still fresh on everyone's mind. It'll freak them out. Go with it if they ask you about it." He smirked.

I shook my head in disbelief and proceeded to change out of my dirty cammis.

Ramadi generated a morbid sense of humor. It was how we endured trials, but the line between humor and lack of empathy eroded.

The following morning during formation, the Platoon

Sergeant called for a volunteer to assist him at the funeral home. The legal process required an individual who knew the deceased to identify the remains before burial. Only a handful of Marines knew Lira. I shamefully believed I didn't know him as well as I should have. In the absence of volunteers, I raised my hand.

The drive to the mortuary echoed with solemn fear. Anticipating the sight of his body began removing the dignity of his memory, knowing it would be the last image I would have of him. His coffin was staged in a dim corner–drawing no attention among an assembly of wooden cells. We were warned of his deteriorated condition and then asked to identify physical features that would confirm his identity.

They thanked us for our time and assured us he would be given reverent care. His timely death before his court martial left him eligible for burial with military honors. At least he would be buried with some kind of dignity. Still, his absence in Ramadi was a disgrace to fellow Marines. I relapsed into apathy, regressing further into a place of darkness. Anger burned inside me as what was left of my empathy poured out for his wife and daughter. They would never have to see him in his final condition.

His struggle was over and wasn't worth becoming my own. If his loss was to mean anything, I had to move forward as I did with the others. I was no good to anyone if I was out of the fight. The war materialized inside me again–anger without release, confusion without consolation. His death was a conduit for casualties from Mesopotamia to the West Coast. I raged against the faculties that imprisoned my mind. I grew distant from others without understanding how to heal, but maintained appearances. I was deteriorating in dignified posture.

The face of defeat could never be displayed in front of the junior Marines. I was now a leader, hardened by grief and the persistence of isolating trauma. Lira was an imminent reminder of complacency at home and the knife in our backs if we failed to fight the enemy within us.

* * *

"PFC Reisberg, my room, 1730hrs." I called to the formation of new Marines.

"Roger Lance Corporal." He replied.

Battalion needed Forward Observers now that the senior FOs were gone. We kept a watch on the new guys to see who might have the potential maturity needed for the billet. We made our decision based on first impressions and obedience to orders. We looked for a maturity that could weather the impending storms.

Reisberg was a few years older than his peer group and a year older than Mike and I. He joined the Marine Corps after a few years of college but decided he'd rather serve than accumulate academic debt. He was perfect for the billet. We needed two replacement FOs and Private (PVT) Rodriguez was our next choice. He was quiet, making his reserved presentation a mark of potential maturity in a group of untamed fledglings.

"Private Rodriguez," I called out to the formation again.

"Here, Lance Corporal." He replied.

"You're coming too, 1730hrs, my room." I ended before dismissing the formation.

Reisberg and Rodriguez would be our responsibility for the rest of our time in Weapons Company. We taught them everything we knew about CFF and the role of a Forward Observer. Once they were able to hold their own during 81mm ranges, we assigned them to a line company.

The flow of our training and time back from deployment

gave me the confidence to step out of my own needs. Stepping into a leadership role forced me to assess my own spiritual health. I arranged time during a handful of field exercises to hold group discussions on God and faith. Close friends gathered to listen and a few curious Marines chimed in with their opinions. I was ill equipped theologically, but moved forward despite the heckling of others.

"Can a person kill someone and still go to heaven?" Someone interrupted. "I'm already screwed. There are only so many that can make it anyway."

I didn't know how to respond to his questions, but I could tell he wasn't there for answers. This particular Marine had a different faith background and held a decent amount of knowledge of Scripture. His disdain for God developed over multiple combat deployments to Iraq. Clifton hijacked the conversation and poured fuel on the tense fire.

"What do you have to say about that Sanderson? Probably something you'd like to know." His contempt reserved no sympathy. Severe trauma and spiritual deprivation were rampant. Demons from Ramadi surfaced to fight in every attempt for spiritual healing. Through the strength of our camaraderie, they infiltrated from within.

Lance Corporal Asher Flynn, an Assaultman from Missouri, joined our weekend crew and began to open up more than I had seen since his time with us in CAAT Red. I invited him to LA and he took up surfing with Cory, Josh and I. He kept to himself for most of the deployment to Ramadi and when asked why he suddenly became social, he stated that he wanted to avoid close relationships in case some of us didn't make it home.

Cory struggled to paddle out as his shoulder required time to heal. Tiny pieces of metal were still lodged in his arm. He scratched at his wound like a mosquito bite, occasionally tearing blisters to free metal shavings. I was disgustingly intrigued.

"Are you serious Cory?" I said curiously.

"There's too many small pieces peppered throughout my shoulder, they only pulled the larger ones. Doctors said this would happen as the body's way of rejecting foreign objects. They're small enough to be pushed out on their own." Cory explained before optimistically shifting from his misfortune.

"At least I don't have to shave like Peterson though!" He said laughing.

"What happened to Peterson?" I asked.

"You remember the flashbang that exploded in his hand on deployment?"

"Yeah, I remember. He lost a few fingers, didn't he?"

"Yes, but he also had to have a skin graft because of the burns. I saw him shaving his hand in the barracks and he said the skin they pulled was growing hair. He has to shave his own butt hair from his palm!"

Cory had the unique gift of resilience and remained sober throughout his entire time in the Marine Corps. Eventually, the trauma would catch up to him as it would the rest of us.

Josh and I also grew closer under trial. He was present for my emotional breakdown before Ramadi, and I was there for his. He developed a minor spasm shortly after our return—evidence of a traumatic brain injury (TBI). On a late-night drive back to base, Josh erupted shortly after we departed LA. I signaled across the first two lanes of traffic after entering the freeway when an unidentified object lay stretched across our lane.

"Look out!" He broke the silence just before rolling over the obstruction.

"What was that?" I said trying to compensate for the blind warning.

"Oh no, oh no…"

"Josh, what's going on? what happened?" I said with pure confusion. He was unresponsive and hyperventilating.

"Josh, talk to me. What did you see?" I pulled to the side of the freeway. He bailed out of the vehicle, stumbling around the shoulder. I took a few breaths trying to understand what went wrong. I was paralyzed in analytical fear; Josh was detached from himself. He climbed back in and quietly explained the outrage.

"I was driving the high back on a night patrol in Ramadi when we came upon wires crossing the street. My VC called out at the last second and I slammed on the brakes. We almost hit a pressure plate IED that would have destroyed the truck. I saw something in the road and I thought it was pressure switch."

"You scared the hell out of me dude." I said relieved yet visibly upset.

"Well, I thought we were dead."

The time came to break from conventional warfare training and form CAAT teams again. We were now running the platoon's day-to-day operations at the NCO level. I was free to resume faith discussions without the disruption of senior Marines just as invisible wounds were beginning to unravel. Weapons Company neared our final months before departure but not everyone was fit for service.

A handful of Marines were exhibiting signs of post-traumatic stress that compromised their ability to perform under pressure. OG's sanity was caving and falling into bouts of depression regarding Conley and Fitzgerald. Most of the Marines in our peer group suppressed their inner demons, but OG was the first to come forward and engage with them. He displayed no physical impairment and was ordered to proceed with training. Cory's mental state was also compromised and was forced to remain behind while 3/7 redeployed. We received word that we would be returning to Iraq at the beginning of April, picking up in the season we had left behind a year prior. We were heading back to Ramadi.

CHAPTER TEN
RECONSTRUCTION

I STEPPED OFF THE RAMP of the CH-47 Chinook unearthing a moon like dust intimately reacquainted with the souls of my feet. The smell of sewage in the spring forecast an authoritative summer roast that would torture our patrols with vengeance in the coming months. Aesthetic changes to Camp Ramadi and Hurricane Point were futile in disrupting previous memories. The sun would rise a thousand times in our terms of service and no amount of light could chase the shadows from Ramadi. The city was a display of unforgiving scars.

I closed my eyes to meet the devil in my mind, fixated on a debt that was owed. I was confined to a conflict of insatiable anger. I masked fear through an erected image, denying the belief that what might come would be worse than death itself. I protested violently within myself. I thought of the junior Marines and snapped back from trailing thoughts. The war in my mind would be everyone else's to fight if I kept giving in. I breathed in the warm pollution, pushing forward with new layers of protection around my heart.

I was ready to use force absent of emotion. Violence was a tool of the trade. There was no utility in entertaining doubt when it was part of the job. Apathy was an adulteress sculpted by emotional tyranny—numbing the bad paved the way for action. The person behind the uniform was suppressed to be what the culture expected. I set an example for the new Marines who wanted furious tenacity.

My eyes dilated as they cracked open, looking past the subtle changes of scenery. I inhaled another breath of JP-8 fumes. My desire for understanding wrestled with the demand for action. The adversarial relationship between impulse and intention burned in conflict. Even the condition of Ramadi shifted like the war between my heart and mind. Weapons Company moved from offense to defense. The chance to reconnect with human values would challenge our posture.

In the summer before our return, Sheikhs and Tribal leaders formed a council to fight against Al-Qaeda. Sheikh Abdul Sattar Abu Risha began the movement known as *Sahawat al-Anbar* (Anbar Awakening), later to become the Anbar Salvation Council. Their campaign conducted raids at night, often using axes on unsuspecting fighters. With the United States backing their efforts, the insurgency began to crumble. Al-Qaeda lost the fight in Baghdad and Fallujah and now they were losing the Anbar province through Ramadi.

Our return to Ramadi was met with a dynamic shift in operations. We sweated over humanitarian aid instead of the lead farming we were used to. Weapons Company's humane role discouraged junior Marines who were looking to play with fire. They were bloodthirsty for a Combat Action Ribbon (CAR) and eager to know what we knew. We were blessed with declining hostility, while at the same time keeping them ready for violence to surface without notice.

The insurgency was losing ground in the expansion of local Iraqi involvement, specifically with the establishment of a new police and military force. Iraqi police officers secured major intersections, discouraging the placement of IEDs. Buildings damaged by gunfire and RPGs were reconstructed, while roads were swept and maintained for the first time since our occupation. Sewage and power lines were repaired and those who lost family members could file a claim for reparations.

True to their deceptive principles, the enemy's hatred was indefatigable. Their persistence resumed by targeting Iraqi forces. In a violent political protest, they detonated a VBIED next to Sheikh Sattar's vehicle on the highway just north of Blue Diamond, killing him and three of his guards. The explosion rocked HP, sending me and a few other Marines running to the hooch for accountability. The disruption inconvenienced Corporal Rosales and I–nearly pummeled by ceiling tiles while on the bench in the gym.

Sporadic attacks threatened Iraq's stability, reminding us there was still a war happening. An adjacent Marine unit lost two Marines and seven others were wounded when a suicide bomber detonated himself near the Habbaniyah Canal. Similarly, another suicide bomber managed to charge the front gate at Camp Ramadi with a VBIED killing a handful of contractors. By that evening, it was televised on CNN. Our work in the Anbar province was broadcast across the world, keeping family and friends in the realm of fear that should have been reserved.

However, the humanitarian context of our mission helped pacify anxiety even with random attacks. The Battalion's order to win the hearts and minds of the Iraqi people was being achieved.

A 5k-community run down MSR Michigan celebrated the work being done despite any threats of violence. Iraqi flags streamed across recently cleaned roads and boosted local morale. Senior ranking Officers made a presence with religious leaders for the celebration and shared in the joy of a recovering city.

To our advantage, half of the Company knew the city's geography. We felt connected to the recovering city, unflinching to all threats. The routes and architectural destruction were memory markers for the pain we spent a year trying to erase. Though at times, we stood in our own way with unhealed wounds. The fear that reigned on our first deployment dissolved. We were incredibly bored. I lost the drive to call home and stay connected with the outside world; secluding myself into a pair of headphones. I was finally present.

Downtime grew heavy as I fought against mediocrity and the thought of Kristi. I blocked her out in order to function, but there were fewer patrols to occupy the majority of my time. There was still a vacancy. There had been no healing, no comfort in the form of love. My heart broke again with the isolating time to finally process a breakup. I underestimated the time it would take to recover. The environment had changed slightly, but the time in between deployments faded off. She was still here, where I left her and where I hoped she would remain. But I came back.

Ramadi was recovering faster than I was. Prayers quickly offered an escape, but felt as if they evaporated just as fast as they were spoken. Music shuffled through my headphones like a self-loathing soundtrack. My pulsing aggression temporarily calmed with the opportunity to serve in a capacity outside of our traditional role. Supplying aid to families across the city resonated at a frequency my heart was desperate for.

The Battalion was given a chance to uphold its own values—
No better friend, no worse enemy. To show hospitality in a city
plagued by violence was a tremendous victory. We sacrificed
through service and force. Still, the calm of Ramadi became a
visceral saw to some wounded minds. We maneuvered down the
same roads on supply runs, where business cards were etched on
the ground. The lack of fighting caused others to wander into
their inner darkness with all the sensory cues to expose scar
tissue. OG's mental state continued to deteriorate beyond other
Marines' as the environment reopened old wounds. His the-
atrical behavior masked his declining mental state.

The winding down of our second deployment brought a
joyful hope to the senior Marines, knowing we were not likely to
make a third deployment. I had no desire to undergo the fatigue
of a third conflict. We were blessed to be on cruise control and
counting the days until our return. Our final flight out of the
Middle East was scheduled to land back in the States from
Kuwait in time for Thanksgiving. Weapons Company staged in
tent city after our RIP with the incoming unit. The communal
space offered zero privacy, but allowed tensions to decompress
before reuniting with families.

"Eight days and a wake-up, ladies!" Corporal Marzillo
triumphed in the large squad bay, "Pretty soon you'll be
hammered on cider and faceplanting the turkey at the dining
room table next to traumatized family members talking about
killing bodies and eating babies."

"Yut!" another Marine concurred.

Marzillo continued with his comedy hour by directing his
attention to a fellow Illinois resident, "Don't worry Reisberg,
you're more likely to get your CAR in Chicago than Ramadi at
this point."

Reisberg shook his head in disapproval while another Marine chimed in, "Y'all are both from Illinois, right?" he said pronouncing the silent "s" in Illinois.

"It's Illinois." Reisberg corrected.

"You pay for the 's', you should use it." he joked.

I spent a great deal of downtime at the smoke pit with Josh when I reached my limit for uneducated humor. We reflected on our first deployment to Ramadi and tried to pass time constructively. On our last night in country, I flipped his last remaining cigarette inside the pack to frustrate his nicotine habit, which supposedly made it bad luck to smoke.

"Did you flip a lucky cigarette? I can't smoke this." Josh accused.

"I'm just helping you quit one at a time," I replied.

"Why won't you smoke it anyway?"

"It's bad luck, I won't do it."

"There's no such thing as luck."

"Listen, I smoked four luckys in Ramadi on our first deployment. Every patrol right after that, I was hit by an IED. It's bad luck!" He exclaimed. "Dude, that was my last one too."

"Josh, we're on a flight out tomorrow. I think we're pretty safe from IEDs at this point." I laughed at his mystical belief and hassled him into submission. He sighed, contemplating a trip across base for a new pack. "Whatever, I'm smoking it." He admitted it would be his last one in country.

The following day, Weapons Company staged gear and eagerly awaited our flights home. Marines stayed in the area for accountability and avoided unnecessary chow hall runs. A Platoon Sergeant addressed the Company with word from the chain of command. "Listen up Gents, there was a problem with the flight schedule. One of the line companies was pushed ahead of us so we're going to be here another day until they sort it out."

A wave of aggravated comments followed before he could finish. "I realize this means we're going to spend Thanksgiving here, so once we're done go ahead and call home to let them know."

Josh directed his attention to me with wide eyes and a matter-of-fact stare, "Smoke the lucky, what's the worst that can happen? It's not like you'll get blown up here." he exclaimed sarcastically. "This is B.S. I need a cigarette."

"I thought that was your last one in country?'" I prodded further as he grabbed his rifle and flipped me the bird while heading out to the smoke pit to bum a smoke.

* * *

We returned to the States having missed Thanksgiving, but we were thankful our feet would be firmly planted in US soil for a while. Shortly after our return, Battalion scheduled Weapons Company for a combined arms training regimen in South Korea. We were redeploying for just under a month, six months before I received discharge papers. Some were excited to train with the Korean Marines, while others continued the trend of communal complaints. I pioneered a few myself.

Attitudes came around once we began movement in the field with a distinguished foreign military. Snow covered the hills and valleys painting a cold weather scene that overlapped sepia tone memories of a scorching desert. We hustled through drastic elevation and temperature. Mike and I supervised Reisberg and Rodriguez throughout the training cycle. The knowledge passed to us from our senior FOs flowed beautifully through two generations of Marines. The intimacy of our small billet allowed for us to personally see the fruits of our labor. Their ability to provide fire support without oversight removed some of the barriers between senior and junior Marines and ushered in humor to break the monotony of downtime.

After Korea, Weapons Company resumed conventional training in 29 Palms. I would skate out my time through the various training exercises and witness the junior Marines become leaders over the next generation of warriors. The intensity of their leadership would carry these new Marines into the next battle–reflecting what they learned from us. My original disdain for tradition turned into appreciation upon seeing their leadership flourish.

The junior Marines were prepared to receive the next drop of Boots, making them senior Marines and giving us terminal status. Our aggression was settling–allowing honest conversations to expel true feelings. Infantry culture prevailed in hardening our spirits, where grave humor never ceased and personal insults were terms of endearment. We were a community of brothers who knew our time together was coming to an end.

I let my guard down anticipating the point of closure for active service. The fear of death and a prolonged posture of hostility faded in preparation for the transition from Marine to civilian. Less than three months remained until my EAS. Knowing I would see life beyond my contract authored a new sense of purpose. The last stretch of training as a seasoned NCO gave me time to consider life in the civilian world and de-institutionalize.

Conversations grew into meaningful connections and a sorrowful hope for our endeavors on the outside. A strange new sense of responsibility took hold. We would have autonomy over our lives and relationships again. The thought was overwhelming for some who had families, making reenlistment an appealing option. Most planned to attend college or begin work in the private sector with the aim of finding a healthier work environment and higher pay.

Long nights of drinking and reminiscing about Iraq fulfilled our remaining days with celebration and depression. I broke even further from my straight-edge lifestyle and drank with them on occasion. Socializing helped unpack the dark baggage we were carrying. We listened to each other's words or simply shared in a mutual silence without a fire pit. The memories of the Marines we lost were more powerful in our time together; the value of their presence gave meaning to our fight.

OG was still struggling in his fight. His demeanor was declining rapidly even though he was finally able to see a specialist. Weapons Company transferred him to another Company to avoid a conflicting schedule with field training and appointments. It appeared beneficial to provide an environment removed from the strain of ranges and ruck marches, but only separated him from friends and a world where he was known.

Night terrors and hallucinations became delusions and pulled him deeper into isolation without accountability to monitor his drinking and medication intake. He tried to hide from the shame of his condition, but his diabolical humor was suspect. Mike and I were in the chow hall when he approached us one afternoon:

"Hey guys, so this might sound weird, but I talked to Conley and Fitzgerald last night. Is that strange? I mean it was probably just a dream, but it was like they were there right? I don't know, crazy."

OG caught both of us off guard in his rather excited tone of voice. It was as if he was pleased to see them again, yet knew the experience was wholly irrational. He understood the fallacy of his vision but was desperately hurting.

"Are you saying you had a dream about them last night?" I replied trying to understand his concern.

"No, I had a conversation with them. They visited for a few hours. In my room, I guess. Have you guys ever had any dreams like that?" He exposed the severity of his condition.

We saw through his casual presentation into something darker.

"I don't know man, sometimes it's hard to sleep. Have you been able to see a doctor about it or is treatment being held up?" I said.

"Yeah, they gave me some pills. I miss being with you guys though. Motor-T sucks and it's a bunch of POGs." He said expecting an enthusiastic response.

The conversation shifted to our EAS dates to offer an optimistic solution. We finished eating and parted ways without understanding his mental state of mind. I knew he was struggling and Mike had been with him during the night of CAAT Black's loss. We were clueless at how to offer any consolation other than communal conversation. Our hearts were wounded as well.

I hustled through the final months of service trying to focus on a future starved for my attention. We attended transition courses and conducted a few field exercises before checking out. I began thinking of the life I was losing. I was ecstatic to begin a new life with autonomy, but also felt at home in the presence of the Marines. My heart was restless. Going home was all I could think about. I surfed every available weekend with Josh and Asher knowing it would be the last time we'd be able to. They would return to landlocked states. I placed my hope in new ambitions after service feeling bulletproof to the soft world on the other side. The Marines preparing for separation felt the same, yet our minds were fractured.

A Sergeant pulled us together to pass word before departing from a short field short Op. Myself and the Marines present were blindsided as he explained that OG had passed away earlier that morning. He was found in his barracks room—his death was ruled an accident due to multiple drug toxicity. Hearts sank in silence, immediately reverent and beyond the defense of doubt. Our war had ended but we kept losing Marines. The drive back to Mainside was concentrated with strenuous remorse and silent gasps for air. The shadows of Ramadi lingered like smoke clouds overhead. I placed so much energy in the momentum of my next life that I failed to see how much we needed each other in that fight.

The front I was putting up would eventually come crashing down. Ramadi would be waiting for me at home. It was easy to displace grievances in the company of brothers. Life carried on regardless of the internal violence devouring our spirits. We lived through worse conditions and the complexity of trial developed greater character through pain. OG's death caught us at a time of healing and lowered guard. A sacred space was violated causing uncertainty in the road ahead.

The final months of each of our contracts descended like dusk to erase our existence. I began the checkout procedure—attaining signatures, returning issued gear, legal paperwork, and clearing medical records. I submitted a terminal leave request and received my discharge papers in July, almost a month prior to a full four years from the date I arrived on the yellow footprints. Though unlike the formal graduation ceremony we received at MCRD, there was no formal procedure for an individual's discharge. The Marine Corps utilized us exactly as we were: government property. Any celebration or ceremony would be for our own pride.

Marines checking out were placed in a separate platoon to allow for administrative tasks to be completed. I woke up at dawn for roll call and proceeded to the Company Office to acquire my last signature. With papers in my possession, I approached Sergeant Cahalan to notify him of my status. It felt as if I had rescinded my biography–pending approval to live again. I had grown accustomed to the institution and taking back my life required authorization.

"What else do I need to do Sergeant?" I asked.

"Nothing, get out of here." He replied.

"What about roll call, do I need to notify the chain of command?"

"Nope, I'll let Gunny know, he'll cross you off the list." He stated.

I was sure it was more complicated than that. The informality, the casual instruction, and the lack of audience left me skeptical.

"Are you sure I can just leave?" I asked a final time.

"What do you want a ceremony? That's it, you're free to go. It's been an honor serving with you Sanderson." He affirmed with an outstretched arm.

I paused realizing it would be the last time I would wear my uniform and a long time before reuniting with the Marines of Weapons Company. For some, it would be the last time I ever saw them. I shook his hand feeling unconquerable, yet incomplete. "You too Sergeant."

Vivid memories played before me on the drive home without ever looking back. I drifted across the high desert as if knowing where to go. I was heading back without a plan and without a purpose. At twenty-two years old, I was autonomous and blessed with the gift of life. I was in no hurry to make commitments for a while. It was already dusk behind the western skyline of LA; the sun fell quicker than the drive home. I chased the dimming light into darkness with confidence of having fulfilled a purpose.

ECP post with an M240B, overlooking the Euphrates River and palm grove to the east along Route Nova.

Inspecting the IED damaged AT-4 rocket from the back of Alpha 4's truck. (Photos by author)

Two curious Iraqi girls approach Alpha Section during a day-patrol in the Arches District.

My view as driver of Alpha 4 in Ramadi's downtown commercial area known as the "Souk." (Photos by author)

One of Weapons Company's Humvees destroyed by an IED. Vehicles often needed to be towed back to base.

Myself and a fellow Marine playing guitar at the fire pit behind CAAT Red's hooch at Hurricane Point on our first tour. (Photos by author)

Weapons Company on a rainy-day patrol in the Warar district. An ominous figure is standing in the background.

A mangled high-back Humvee is towed back to Hurricane Point in front of the iconic 7-story building at Checkpoint 295. (Photos by Gerardo Rosales)

Lance Corporal Ken Rick kneels with his M249 SAW to greet a curious Iraqi girl on the Battalion's second tour to Ramadi, Iraq.

Mike and I heading back from a field Op in 29 Palms, CA after an 81mm mortar range. (Photos by author)

Shane Swanberg after SOI graduation. He spent some time with 1ˢᵗ Tank Battalion before coming to 3/7 Weapons Co. (Photo by Linda Swanberg)

Iraqi kids hassle Marines for candy and soccer balls. (Photo by author)

Weapons Company patrol north of Checkpoint 296 in the Shirikah district near the Ramadi Mosque, formerly known as Saddam's Mosque.

Chad Oligschlaeger (OG) taking a break from a grueling hike during Mountain Warfare Training in Bridgeport, CA. (Photos by author)

A shot of me staged at the clearing barrels before a mission mid second deployment. (Photo by Eric P. Goodspeed)

Marines conducting a foot patrol through the Arches District to supervise Ramadi's reconstruction on our second deployment. (Photo by author)

A group of Iraqi women walk to school in the Warar district. American and Iraqi forces were unavoidable along the MSR.

(From Left) Myself, Mike, Asher, and Josh the night of our return from our second tour. (Photos by author)

PART II

CHAPTER ELEVEN
THE UNKNOWN

Happy is he who has the pure truth in him.
He will regret no sacrifice that keeps it.
　　– Johann Wolfgang Von Goethe

HOW DID I GET HERE? It has been nearly fifteen years since I discharged from the Marine Corps yet this question still haunts me in different seasons. After the military, a significant way of life ended and a new one began. At first, I felt free. Free to pursue a vocation in a limitless world. Strangely, my reality became the opposite of freedom. In fact, I felt more disconnected from the world than ever. The new question became, *"Who am I now and what am I doing?"* I wasn't afraid of my future but my past. I was in a state of cognitive dissonance, especially as a Christian. How was I supposed to live with what happened? How had my deployments changed me?

Conflict raged within me. Now, I understand that the conflict of faith I experienced resulted from the displacement of three foundational pillars: Identity, Belonging, and Purpose. The Marine Corps was the first vocation in which I embodied these fundamental pillars. I grew up in church, studied Scripture, and knew the "rules" from a religious aspect yet didn't understand how grounding a relationship with Christ was until a year before I joined the Marine Corps. All three pillars are found in Him. My alignment with Christ before the military guided my spiritual development. I discovered adventure, established deep relationships, and found significant meaning in the midst of extreme danger.

Unfortunately, I felt that I had lost all these traits in the civilian world after the Corps. The feelings of isolation and meaninglessness were paralyzing. I doubted my future, questioned my recovery, and lost sight of my purpose. The person I had become felt unwelcomed in this new context of civilian life. People looked at me different. Even within the first hour of returning home from my first deployment my own mother told me that I physically looked different to her. It was as if she could see the pain in my spirit through physical appearance alone. People I came to know in social circles had similar reactions, like they knew something about me that I didn't.

It is both fascinating and scary to see how our communities shape our image. If we are not careful they will ultimately define us. Still, I felt I knew things about life they would never understand. The political climate at the time had shifted under a new presidency and the outcry against the war in Iraq was commonplace. I took pride in my service. I felt confident in my ability to handle adversity. Being told that what I fought for was unjust, that we were dying for capitalist interests, and that God was not behind us was a fight I was not prepared for. It was a factor that further alienated me from the society. Finding a place to belong felt hopeless. I was faced with the decision of either standing up for what I believed at the expense of community, or become like everyone at the expense of who I'd become.

Faith has a way of confronting adversity and bringing us

closer to the people God created us to be. The course of our lives is determined by choosing to identify with God. Scripture says we are made in the image of God (Identity: Genesis 1:27), we are designed to function in community (Belonging: Genesis 2:18), and are created to bring glory to God (Purpose: Isaiah 43:7). These are our foundational Pillars. Fulfillment only comes when we live within the created order. The Fall in Genesis 3 illustrates the disconnect between humanity and God. Yet in his great love for us he gave us Christ so that we might be reconciled to him again in the intended created order.

My biggest struggles through the post-traumatic growth process have been the displacement of my identity, place of belonging, and purpose. The spiritual enemy uses our worldly desires against us. He knows we long to belong and that we seek meaning in what we do. I placed my worldly status above my foundational pillars, becoming defined by titles and not by Christ. I loved my accomplishments more than others. Without an eternal foundation we descend further into false identities, broken communities, and lose our reason for living. The path toward wholeness must begin and end with Christ.

IDENTITY

So God created man in his own image, in the image of God he created him; male and female he created them.

— Genesis 1:27

Identifying with God does not come with the promise of comfort; it comes with an understanding of his character. God designed us in his image and likeness. When we desire him, he will guide us by first showing us who we are. My faith led me to join the Marine Corps after high school. I knew it would not be easy. I walked into the recruiter's office and asked for an

Infantry contract the year we invaded Iraq. That's exactly where I wanted to go even though I knew little of the process. I was terrified to step into an uncertain future, but the confidence of my decision was rooted in faith not my physical security. I stepped into the unknown with conviction and more importantly, an identity.

Earning the title of Marine is an accomplishment I'll never forget but it is not *who* I am. Marching out on the Parade deck at MCRD in San Diego on November 12th, 2004 in front of family and friends after three months of sweat, blood, and tears affirmed my decision and my faith. Looking at the Eagle, Globe, and Anchor in my hand was more than an achievement. It reflected just a portion of who I was becoming. Harder days were still ahead but that moment was a chance to reflect on the work God was doing in me. My ultimate identity rests in Christ even as I took on a title that would last a lifetime.

When I first learned our unit would be deploying to Ramadi, Iraq at the height of the war my faith shifted from a bold idea to an unpleasant reality. God required courage to move forward with my commitment to him despite fear and insecurity. Faith didn't make me fearless, but it did help me to focus my fear on God. God revealed himself to me as I pursued him throughout times of uncertainty. What I didn't expect was an unraveling of more pain in addition to the title. One of the most challenging aspects of life is maintaining the worldly identities that define us. Any title we earn in addition to our eternal identity is temporal.

"Marine" was a title I earned and will always have as long as I live. But it is also something I must keep earning each day. Once a Marine, always a Marine as they say. Similarly, calling myself a Christian means following God's commands in my daily life. The difference being that my eternal identity was something to be realized. It is not something I can lose. We all have an eternal identity as we are made in the image of God. There is nothing we can do to change that. Our actions will either bring glory or disgrace to his image. Becoming a Marine was something I earned and how I act in light of that title defines what it means to others. Unfortunately, many people have painted both titles in a negative light, including myself at times.

I faced an identity crisis after the military as the civilian world rejected what I'd become. Many struggled to reconcile how I could be a Marine and a Christian. The legacy of violence appalled people. Additionally, I felt convicted for swearing so frequently and grew frustrated with the work ethic of others. I was short tempered and easily agitated over the slightest inconveniences. I offended people's sensibilities and challenged their perception of a peaceful Jesus. I grew angry with God and the culture in front of me. His silence felt like betrayal. I struggled to navigate this isolation because my identity was attached to a community that was no longer there. As I poured myself further into Scripture to understand Christ, he began to remind me of my eternal identity.

The Gospel of Jesus according to Matthew, Mark, Luke, and John hold a significant pattern of stories about who Jesus is. Aside from the first chapters of Matthew, Luke, and John, the first 4 books of the New Testament capture Jesus' ministry in the last 3 years of his life. His life before this time is summed up in Luke 1:80, "And the child grew and became strong in spirit, and he was in the wilderness until the day of his public appearance to Israel." Jesus enters the scene after 30 years in a "wilderness." Note Luke's use of the word "wilderness" here. The adversity of each wilderness varies depending on God's purpose, but the function of each period remained the same: to develop intimacy with God.

There are few accounts of Jesus' life before his public ministry. Nearly everything Christ is known for is defined by what happened in a 3-year period. During this period of his life he performed wonders while at the same time offending the religious elite for not appealing to their expectations. They considered him a radical and killed him for it. His life as a carpenter before this did nothing to disrupt their lives. He lived relatively at peace. However, living as a humble carpenter was a temporary vocation and the training ground for his ultimate purpose. Jesus knew his true identity and stepped into a public ministry after his time in the wilderness beforehand. Christ displayed his identity to the point of death and changed the course of history. Here is why it matters:

1) Jesus knew his identity and was uncompromising in it.
2) His authenticity generated community and hostility.
3) He fulfilled God's purpose not society's expectations.

Jesus' life is the standard for the three pillars of identity, belonging, and purpose. Everything he did is aligned with God's will. Thirty years in the wilderness prepared him for his public ministry and the ultimate sacrifice he would make within God's purpose for his life on earth. We will always find ourselves disconnected when we establish our pillars in a seasonal assignment. Regardless of the titles we earn, we must focus on the person God designed us to be.

John the Baptist lived in the wilderness, wore clothes made from camel hair, and ate locust and wild honey to distinguish himself from the religious elite. His lifestyle gave him a better understanding of God's will. John knew his pillars were not found within an institution. The wilderness I found myself in upon returning home began with a fractured identity. My image–

my identity—was saturated in the Marine Corps. My tribe was Weapons Company and my purpose was service. I allowed my seasonal assignment to define who I was. My title as a Marine will always be second to my eternal identity in Christ. Wholeness cannot be found outside of God's design. It is through a relationship with Christ that we discover who we are.

Opposition always seeks to keep us from realizing our true identity because an understanding of who we are illuminates our purpose. John chapter 6 shows that most of Jesus' disciples abandoned him due to the difficulty of his ministry. He was left with twelve disciples; one who betrayed him, and the rest who fled to save their own lives rather than be associated with him at his death. Jesus lost his community, yet was uncompromising in his purpose to the point of carrying his cross in utter loneliness. He was confident through suffering even when the world rejected and doubted him. The enemy always tries to isolate us so we question who we are and what we stand for. In the most desperate moments of his death Jesus declares, "Father, forgive them, for they do not know what they are doing." (Luke 23:34).

The pillars of our eternal design are often rejected by society because they undermine secular values. Our pillars reflect our relationships. For Christians, the enemy will always attack our identity first. The world will always attempt to define us by its own standards. But there is clarity and conviction in the freedom God offers. I am unapologetically a Marine. I am also un-wavering in my faith regardless of the trouble people have in reconciling the two.

I knew serving in the Marine Corps was where God led me to be at eighteen years old. I had no plans beyond those four years to reenlist. And after those four years were up, I struggled with who I was. Living beyond the fulfillment of my military service was a catastrophic pillar check. I fixated on my status as a Marine during that season and lost sight of my foundation. The result was an anxious disconnection in my relationship with others and my purpose. The question, *"How did I get here?"* haunted me for years after the military. Seeking out community after the Marine Corps has been a painful process. Relating to others who don't understand, or even outright reject military service, can often feel like a personal rejection. In the eyes of the world, we are what we do.

Once we understand who we are we can reach a place of belonging. My title and tribe affirmed me in the Marine Corps, but alignment with my eternal design is the only thing that will bring me lasting peace. Any identity outside of Christ will fail and will lead to fractured relationships. The work we do and the people we surround ourselves with shape the way we view ourselves whether we know it or not. The cultures we are a part of shape our identities. If we are not careful to ground ourselves in who God tells us we are, then we will find ourselves disconnected from community and rudderless in our pursuits.

BELONGING

> *Then the LORD God said, "It is not good that the man should be alone; I will make him a helper fit for him."*

> – Genesis 2:18

We are designed to be relational. When I arrived in Iraq and saw the volatility of the conflict, I was more committed to the men around me than the ideology that brought us there. Politics were irrelevant. The cause we fought for had nothing to do with the temperament in Washington. We all had our reasons for being there. Communal suffering became the backbone that built meaningful relationships. I often miss the experiences we had overseas regardless of the pain. Most of the Marines I served with feel the same. When people come together with a shared identity and purpose they form a tribe.

Relationships help us withstand the most detestable conditions imaginable. Chaos strips off the shallow barriers and complimentary laughs that people pay to mediocre entertainment. Every challenge provides the opportunity to see our real selves. We are forced to reveal our character under the pressure that changes us. In the Marine Corps, some of the loudest voices became the quiet ones upon return. The same held true for other veterans I met after the military. Those who inflated their deployment stories were often telegraphing a disconnect from their military family. Those pierced by extreme grief are most affected when they are isolated from others who understand their struggle. A community without shared values and identity can feel the same as being stranded on an island alone.

Our experiences shape us, even grant us respected titles, but they should never become our pillars. Consider our Vietnam veterans who returned home and were largely rejected by society. Withholding acceptance can leave deeper wounds than a traumatic experience. The worst punishment in prison besides the death penalty is solitary confinement. Part of the post-traumatic struggle is separating the pride of service with the trauma involved. Meaningful sacrifice shapes our image. A casual or intentional disregard toward a person's experience is often perceived as an assault on their personal identity. That is why some people would rather align with an unhealthy community than stand with principles in isolation. Still, some seek healing outside of all community.

Christopher McCandless, though not a veteran, is a rare example. Upon graduating college in 1990, he left everything he knew to travel alone and live off the land in North America. McCandless lived an itinerant lifestyle that led to his death in 1992. He wrote passionately about the freedom he discovered outside the tyranny of civilization. He made his home in isolation, yet his soul never found rest. His body was found in an abandoned bus along the Sushana River in Alaska where some of his last prominent words written in his journal captured the emptiness of experiences unshared. We are created to be in community.

The Marines I served with are some of the closest relationships I have. After the military I felt hollow without a close community. In the absence of their presence I made the mistake of allowing the past to become a place of refuge. I sought out those who knew the old me. Friendships that served well in high school moved on with marriages, kids, and relocation. I drove through familiar places in my hometown town to piece together the identity I'd lost. The political climate also changed. Rejection of the war further confused my ability to accept who I'd become. The increasing isolation I felt led me to believe I could find healing through a relationship.

There was a gal I went to high school with who I came to know as a Marine. We lost contact but a few years after the military we reconnected and began dating again. We shared the same hometown, school, and mutual friends. She understood who I was in high school, as a Marine, and now as a Veteran in a way that no one else would ever be able to. She became the conduit for my source of healing. When my parents moved, she was the last intimate contact I had with home. My identity and belonging rested unknowingly on her shoulders. As the relationship dwindled, I felt myself choosing between her and God. I was losing more than just her. Misalignment with my pillars moved me to a place of questioning my source of wholeness. People are a conduit for healing but they should never become the exclusive source of it.

The Israelites were promised a land where they would live in abundance and at peace. They had never seen it or known anything like it before. The journey to get there was arduous and doubtful even though they had experienced miraculous signs and wonders from God who defied reality. They wrestled with the idea of returning to Egypt or continuing into the unknown with faith that God would fulfill his promises. In Exodus 16:3 the Israelites grumbled saying, "Would that we had died by the hand of the LORD in the land of Egypt, when we sat by the meat pots and ate bread to the full, for you have brought us out into this wilderness to kill this whole assembly with hunger." As humans, we prefer the devil we know over the devil we don't know. People will choose toxic communities over no community at all.

Fear of the unknown is one of the greatest human fears. In every journey there comes a point of no return. The Israelites had to cross the Red Sea as part of their journey out of Egypt. The miraculous separating of the Red Sea wasn't just a display of authority to strengthen their faith and provide a way. It also served as a point of no return and a commitment test for the Israelites. After they became disillusioned with God's miraculous works in the desert and expressed the desire to return to Egypt, it was too late for Israel. There was no crossing back into bondage.

Familiarity is a comfortable illusion when the journey becomes difficult. God knew the Israelites would exchange their current complaints for new ones. After 40 years in the wilderness they crossed the Jordan River into the promised land. In Joshua 4:4-7, God commands Joshua to send twelve men back into the Jordan and carry out twelve stones, one for each Israelite tribe as a memorial for future generations so that they would not forget what God had done for them. He was fulfilling a promise that would root them in their true identity and community as God's people.

My fear of an unknown future led me back to the place of known comfort because my new purpose was unclear. When we place our faith in God to fight our battles we often find clarity in our direction. New relationships will be part of this process and old ones may be cut off. When I lost my last anchor of home there was no going back. Nostalgia is a psychological rock from the Jordan that reminds us how we got to where we are. It is not a place of residency. The comfort we have in this life will always be temporary. Where we belong rests on the promises of God which always bring us into community with a purpose.

PURPOSE

Everyone who is called by my name, whom I created for my glory, whom I formed and made.

— Isaiah 43:7

The great commission God has for his people is to make disciples of all nations. The underlying purpose of our lives in every venture is to be a conduit for others to find and follow Christ. God called me to serve at a crucial moment in history. I watched the Twin Towers fall on September 11th, 2001. Two years later I signed a contract for the Infantry. I wanted to be a force for good to ensure that kind of atrocity would be prevented. I was guaranteed conflict overseas. The course was uncertain, but my purpose was clear. I stepped into an unknown future knowing that I would be a light to others throughout the difficulties ahead.

The Marine Corps was a unique place for God to display his strength. That strength included the wisdom needed to navigate the difficulties I would endure. But somewhere along the journey I lost my way. I trusted my own strength and spent less time in prayer. I lost confidence in his will by giving into fear and anxiety. Civilian life became a wilderness of reestablishing trust in the one who kept me alive. My pride was slowly stripped away when I hung up my uniform and stepped onto a college campus. At the start of my second year, I faced a crisis of purpose–an overwhelming sense of meaninglessness.

I scrolled through social media one morning and came across images of fellow Marines I served with. They were having kids, becoming CEOs, and two became fighter pilots. Yet, here I was wasting my life in a basic algebra class with students I would never share anything significant with. One moment I'm combating a hostile insurgency and the next I was nothing but a grey man in the crowd. I had no idea what I wanted to do or where I was going. I felt as if I were being pulled in and out of a desert, not realizing how transitions are the training ground to prepare for the next journey.

Moses spent forty years as a deserter before God brought him back to Egypt to free his people. Moses first had to develop intimacy with God before he could step into his purpose. He performed wonders in Egypt. Then, he spent another forty years in the wilderness so the Israelites could reflect the character of God before inheriting the promised land. Moses was called to suffer with the Israelites in the second forty years. The reasons for a wilderness will vary, but the result is always a strengthened relationship with the Lord.

The prophet Elijah's journey is similar. Elijah advised King Ahab that a drought would come upon the land of Samaria until he declared it over. Then, Elijah fled God's command and hid in Cherith. It is in Cherith that Elijah is isolated from the world in order to understand what God is doing through him. The wilderness prepared him with the character and confidence needed for his public actions. From Cherith, Elijah returns to Samaria where he brings fire down from heaven, destroying the prophets of false gods, thus reconnecting Israel with the one true God of heaven. When King Ahab's wife Jezebel threatened to kill Elijah, he fled back into him the wilderness again.

The wilderness is where God reveals his character and his purpose for our lives. It was in the dull moments and isolation that I began to see God's purpose in what seemed like chaos. Confusion naturally came with new territory. I experienced the loss of home and a relationship during my first years after the military. Isolation taught me to rely solely on my relationship with him in Iraq. Now, I was being led to do the same at home. As I grew in my faith, God led me to other people. True relationship with others is grounded in the character of Christ.

Throughout the book of John, the author refers to himself as "the one Jesus loved." John's identity was rooted in the fact that he mattered to Jesus. Jesus even renamed three of the twelve disciples as a way of revealing their purpose through their identity. The same is true in the Old Testament where God renames those who will bring about his purpose. Jesus spent a significant amount of time building up his disciples to follow his character. After his death, the Disciples returned to their normal lives because their pillars were rooted in the physical Jesus. Peter was especially distraught from denying Jesus three times before his death. Peter, being one of the disciples Jesus renamed, returned to his life as Simon the fisherman.

The Disciples realized their identity, purpose, and belonging were rooted in an eternal, living God upon Jesus' resurrection. One of the most beautiful moments after the resurrection is when Jesus pulls out three declarations of Peter's love for him, reestablishing his pillars. Three times Jesus calls Peter "Simon," shedding light on his identity disconnected from Christ. These three declarations from Peter denounce his three denials prior to Jesus' death and help him reclaim his identity and purpose. Jesus then tells him to "Feed my sheep," signifying that he would continue the work God started in him.

The wilderness is a place of development, but it can also be a place of desolation if we are not careful. Repeated seasons of isolation and feeling cut off from our purpose is a sign that we are disconnected from God. Faith begins by trusting God is who he says he is. Paul tells us that God can always be trusted, and that our suffering is nothing compared to the glory that will be revealed. God showed us who he is so we can understand who we are. Each wilderness we face is developing the character we need in order to carry out God's purpose.

CHAPTER TWELVE
RELATIONSHIP VS RELIGION

When you see a man with a great deal of religion displayed in his shop window, you may depend upon it he keeps a very small stock of it within.

     – Charles H. Spurgeon

I WAS ASKED, "Why would you tattoo your body, aren't you a Christian?" by most Baby Boomers who saw my fresh USMC tattoo at eighteen years old. The question was typically followed by a passage from Leviticus. I was half attentive to the religious ideology behind their opinions. I disagreed, though I wasn't theologically versed in why I did. It resonated as overly religious. At the time I found a few shirts that stubbornly confronted their opinions saying, "Jesus loves me and my tattoos" and "It's against my relationship to have a religion." It sounds ridiculous now, but that was the Christian youth culture when I was in the Marine Corps. Older generations clung to traditions while a radically changing younger generation sought relatability.

Everything I learned growing up in church started to make sense when Jesus became personal to me. The rules and behaviors associated with being a Christian became relational. The Christian God of religious expectations became a relational

God who was revealing something more. The Bible was not a book of rules but a map of how faithfulness would lead to fulfillment. I spent every week meeting with a small group, seeing Jesus through a new lens. It was an amazing revelation of what his life meant then and what it means in modern times. As my relationship with Christ grew so did the strength of my pillars.

I stepped into my first assignment when I received the title of Marine. The first thing I did was tattoo it into permanence. Many scolded me but they weren't considering that my life was likely to be cut short with where I was heading. "Permanent" had a short-term meaning. I felt confident in my calling because I had postured myself toward God. My relationship with Christ cultivated new expressions of faith. For example, my USMC tattoo includes Psalm 91:7 and John 15:13. Regardless of the Christian influence behind it, some rejected the tattoo because it didn't align with their religious tradition. Tattoos were seen as a sin rather than an expression of a life-changing relationship. This was my first collision between religion and culture.

By culture I mean the values, norms, societal beliefs, or practices outside Christianity. Some cultural expressions are sinful while others are cautioned against. However, the comments I received about my tattoos were less about disobeying Jewish ceremonial law in Leviticus 19:28 as they were about a demand for submission to a preferred tradition. Some churches formalize extrabiblical practices to reject cultural influences like this. Yet, churches that use regulations as a means to suppress culture are creating a new culture of their own through denominational constraints. There is a reverence to be appreciated in these rules, but also the threat of compromising theology through legalism and a works-based system.

After the Marine Corps, I faced a new conflict within the

Christian community–an overwhelming emphasis on the experience of faith. This phenomenon came to be known as progressive Christianity. Also known as liberal Christianity, liberalism, progressivism, or liberal theology, this movement grew within evangelical circles following the Reformation in the 16th century. Progressive Christianity is a domesticated version Christianity which appeals to the sensibilities and bows to culture. Cultural demands for the *feeling* of spirituality became the primary goal of the movement. The quest for a religious experience plagued the faith as pluralistic. The movement is an attempt to make Christ palatable to others through a vain concern with relevancy to culture.

Both extrabiblical dogma and cultural influence lead to a deviation from theological doctrine if they are not tempered by Scripture. Proverbs 4:23 says, "Keep your heart with all vigilance, for from it flow the springs of life." Acting on the core principles of Jesus with a genuine heart for his Word will bring out the fruit the world needs. There will always be a balance of religious constraint and cultural relevance within Christianity. If religion is our primary influence, we will work for our salvation through traditions and rituals. If culture is our primary influence, we will seek sensual experiences. But if Christ is our ultimate influence, then Scripture will guide us as we navigate the influences of religion and culture to produce fruit that reflects the character of God and ushers true joy into the world.

CHRISTIAN LEGALISM

> *Obedience that flows out of gratitude is the only obedience acceptable to God and is the only obedience that will bring joy to our own hearts.*

> – Jerry Bridges

I remember waking up early on Sunday mornings for church with my family. My father wore a suit and tie while I was pestered about keeping my shirt tucked in. What I most looked forward to were the boxes of donuts available when we arrived. We had a predictable routine every week. Sunday School started at 7:30am before the main sermon. After church, we would go to lunch with other families then head home so my parents could rest. There was nothing overtly religious about the way we gathered on Sunday, but as I grew older the disparity between our church's beliefs and culture became clear.

In high school I began questioning how our denominational practices originated from Scripture. My mother grew up Catholic, so I learned early on that not all denominations are the same even though we believe in the same God. But if we differ in our practices then which church is "doing it the right way?" Were suits and ties required on Sunday? Does the Bible instruct us not to dance or watch movies? Are we never to touch nicotine or alcohol? All of these questions originated from observing denominational constraints. These practices sought to keep us from the temptation to sin yet often clashed with daily culture. There was no wonder why I was met with protest when I received my first tattoo.

Religious constraints are often a response to cultural influences which threaten the body of Christ. The intention of these rules is to keep people unstained from the world as James 1:27 says. Yet, many ills have been carried out in the name of God through extrabiblical or "non-canonical" practices. Christianity is stained by horrific events that are a result of non-canonical behavior. Man-made regulations create barriers between Christians and confusion toward the Word of God when they are unsound and persist longer than the cultural influences they were erected to combat. A legalistic church conditions Christians to adopt a works-based faith by focusing on behavior modification instead of our need for God's saving grace.

Catholic and Protestant are the two primary denominations of Christianity. Denominations have a history of adding to the gospel of Christ to the point of generating entirely new denominations. How are we to know which is the *right* one? One thing that theologically grounds each denomination within Christianity is the act of worship. The most common forms of worship are seen in the sacraments of baptism and communion. Regardless of the various hierarchies, doctrines, and traditions the sacraments are universal to Christianity. Worship or "praise" through musical expression however, differs greatly in practice. Historically, musical expression that reflected culture became a point of contention.

The early Christian church excluded the use of instruments due to secular and pagan cultural influences. Instruments were widely associated with negative spiritual influence and as such were banned from use in the church. Eventually, organs, pianos, and stringed instruments made their way into musical services while churches maintained bans on others. Most Christians take no issue with instrumentation and agree with reducing harmful influences of secular culture not the tools themselves. The proclivity to focus on inanimate objects, non-essential doctrines, and behavior modification rather than the character of Christ is the mark of religious legalism.

The danger of non-canonical practices and beliefs is that they can be weaponized when they become doctrine. One of the most profound examples of this is seen in the Catholic Church's denial of a heliocentric solar system. In the 17th century, the Catholic church condemned Italian astronomer Galileo for claiming that the sun was the center of our solar system (heliocentric) which conflicted with the widely held belief at the time that the earth was actually the center (geocentric) and that the sun and moon revolved around us. Not only did it take the Vatican over 350 years to formally admit their mistake, but they also conducted a 13-year investigation to reach that conclusion in 1992. To the Catholic Church's credit, Galileo also claimed that the sun was the center of the universe, not just our solar system. So, he was mostly right.

The point I'm making here is that when a religious institution mandates certain creeds, beliefs, and practices they become a governmental body of human authority. They inevitably create a legalistic culture Christians should avoid. The Bible does not give a definitive answer on the centricity of our solar system but allows us to discover truths about how every created thing exists in conjunction with the Biblical narrative. A certain belief about the solar system is a peripheral issue to Christianity. Placing Galileo under house arrest and suppressing a contrary worldview with the threat of capital punishment is theological misalignment. The Bible tells us the *who* and *why*, not always the *how* and *when*. We can always disagree on peripherals but should not resort to violence over extrabiblical violations.

Denominations are overzealous for extrabiblical tradition. We should guard against allowing religious doctrines to deepen the chasm of faith by appealing to traditions over Scripture. Christianity and culture will always be opposing forces but the balance of both allow for a beautiful expression of faith. We are called to live for Christ, but must not fall into the trap of believing we can earn our salvation. Too many rules will suffocate our relationship with Christ. Religious constraints historically lead to tyranny. Similarly, not having a disciplined faith will also lead to failure. Scripture shows us what obedience looks like and to not be taken captive by the allure of culture.

PROGRESSIVE CHRISTIANITY

Liberalism is not another denomination or any kind of legitimate option within Christianity. Rather, it is another religion.

– D.A. Carson

A friend of mine referred me to a new church after returning home from my first deployment. It was non-denominational and practiced baptism and communion. All other forms of worship reflected modern culture. The church met in the auditorium of an old college university. The first service I attended had a full band for worship, an interpretive dance performance, and an artist painting in the corner during the message. Up to this point, I had only known the protestant church I grew up in and similar denominational bodies. I had never seen anything like it and I connected instantly.

I loved the artistic expression and freedom from religious constraints. It was equally freeing to see creativity flourish on weekends home from regimented field training. The church accepted anyone who walked through the doors, including me and my tattoos. Having only been back from Iraq a few months and dealing with an existential burden, this church became a place of refuge. I spent every Sunday home on weekends going to a service. I brought other Marines with me who also enjoyed the radical experience of Christianity. Some of them did not grow up in a church but were open to attend as it shed the burden of rigid tradition. I found a place to belong and spent more than a decade serving there.

Over the years I began to dig deeper into theology. Part of it was from a genuine desire to understand Jesus better while the larger part was to find wholeness after the Marine Corps. This

172

church played a major role in my relationship with others and eased the restraints of organizational control. I went to bible studies during the week and connected with a lot of great people, which significantly helped in my transition and recovery. Then a major shift took place.

The church relocated to the heart of Los Angeles where the cultural narrative permeated the Christian message. The theological truths of the gospel slowly became private, giving way to preferred topics of discussion in the public realm. The church held a policy not to teach from the book of Revelation in any primary service. It took on a prosperity tone and avoided culturally sensitive issues on which Scripture is definitive. Over the course of a few years, the church had seen exponential growth yet transformed into something different from when I had first entered. So, what exactly happened?

The church pioneered its way into a Godless culture with impressive results. This was a huge victory compared to most organized denominations. Their relevance to culture had an observable effect. It was widely accepting of all people which opened the door to those skeptical of Christianity. In redefining the experience of church, it isolated the Apostle Paul out of context in 1 Corinthians 9:19-21:

> "For though I am free from all, I have made myself a servant to all, that I might win more of them. To the Jews I became as a Jew, in order to win Jews. To those under the law I became as one under the law (though not being myself under the law) that I might win those under the law. To those outside the law I became as one outside the law (not being outside the law of God but under the law of Christ) that I might win those outside the law."

In their efforts to spread the message of Christ in the modern age the church advocated for a seeker friendly environment rather than biblical literacy. The church caved to the pressures of political correctness and cultural relevancy. The influence of culture reduced the church into pluralism. It became an inverted version of its conservative denominational counterparts and a proponent of progressive Christianity. The church's theological foundations were compromised for the sake of creativity over structure. People from all over the city were attending a church for the first time in their lives. The momentum of growth made it difficult to see the shortcomings in which Christ was not being preached.

Growth became an idolatrous pursuit in the seeker friendly environment where attendance was the metric for success. People who had no experience of God felt comfortable with the *idea* of Christianity even if they did not profess to be a Christian. Some attended the church for years while maintaining Atheist, Agnostic, Buddhist, and other religious views while benefitting from a community that embraced them "just as they were." Cultural influence shaved off the abrasive edges of Christianity, causing people to adopt only the Christian principles that aligned with their more dominant worldviews. The passivity of the church resulted in a safe space where people could remain in their sin under the banner of Christ.

Progressivism within a church highlights the positive tenets of God's Word while avoiding the painful ones. This hedonistic presentation of Scripture focuses on an individual's experience. The result is a feeling-based faith grounded in the desirable characteristics of Christ at the avoidance of recognizing his inevitable judgement. It provides a comfortable place to belong so long as life remains soft and unchanging. This theology is weak at developing Christians who can withstand the perils of tragedy. In the end, people are left with a shattered worldview when they experience true suffering.

Progressive Christianity explicitly leaves out essential doctrine in order to make church palatable. One of the most obvious indicators of this destructive movement is when the church avoided teaching about sin and judgment. The pastor focused only on the grace of Christ, which diminishes his divinity with God and warps him into a metaphysical genie who grants wishes and takes away pain. Grace is a major part of the gospel, but without a healthy fear of God and the purpose of his wrath we end up violating 2 Corinthians 6:1 in which Paul tells us not to receive God's grace in vain.

The conviction of sin is an essential doctrine within Christianity. Without it we cannot move toward wholeness and, ironically, cannot fully realize our need for grace. Destructive ideologies will creep into a congregation if people remain biblically illiterate. The progressive church I attended found creative ways to side-step sensitive issues rather than risk offending people with the truth. Christianity is not a passive religion. It is confrontational because it does not tolerate sin. An affinity for cultural expression will inevitably lead to a retreat from theology.

The balance of Christ and culture starts with an uncomp-romising stance on theological truths. God is able to work through our lack of understanding as we mature in faith. The appeal to culture in my post-military church initially promised a place of belonging even though it was misaligned theologically. Eventually, I began to feel disconnected despite being sur-rounded by people because my pillar of belonging was rooted in the culture, not Christ. We are not called to make Christ relevant to culture but to reveal Christ to our culture. The church was established as a means of leading others to Christ-like trans-

175

formation through relationship. The movement of Progressive Christianity is so overly sentimental in its approach to spread the gospel that it has effectively become pluralistic. It is nothing more than universalism with a Christian Savior.

CHRIST AND CULTURE

See to it that no one takes you captive by philosophy and empty deceit, according to human tradition, according to the elemental spirits of the world, and not according to Christ.

– Colossians 2:8

Obedience to God's commands was never meant to be transactional; nor can tolerating sin under the guise of relevance to culture be considered obedient. We are known by what we love. If we love Christ, we will embody the fruits of the spirit. It is the fruit of the spirit that convicts us of the proper response. Paul writes in Galatians 6:15 that what matters is whether we have been changed into a new creation. There is a freedom from religious constraint that came after the fulfillment of Jewish civil and ceremonial law. The law of grace came as a result of Christ's gift of salvation. It is the work of the Holy Spirit, not ourselves, that saves and the Word of God which calibrates our moral compass. Our conflict between Christ and culture has become a battle between legalities and sensibilities. This is where Scripture informs us on matters of conscience.

My first experience with a conflict of conscience came in elementary school. I was told a story about a man who received exactly one cent in excess change from a sales transaction. He felt so convicted about the money that he turned his car around and returned the penny. This story is one of religious legalism in which obedience is required for justification. Keeping the penny would not have violated any civil laws, yet it was told as an example of Christian obedience through a violated conscience. A similar and more serious example would be infidelity–though highly immoral, it is not illegal.

The opposite of this type of conflict of conscience is when an action is illegal but not immoral. Speeding while driving is a safety issue. The consequence is often a ticket. However, a person's conscience isn't violated by any real measure if they are only a few mph over the limit. Additionally, a ticket given for this violation is more about obedience to authority than it is a moral violation. Moral law is not violated, though civil law has been. As frustrating as it may be God does not promise immunity from civil repercussion. These violations can be frustrating matters of conscience for a Christian.

Matters of conscience are especially present within the military where submission to authority is institutionalized and life and death scenarios are common. In 2009, Marine Corporal Dakota Meyer disregarded his command's order to remain on base during a firefight and entered into enemy territory to recover wounded Marines and Afghan troops. Corporal Meyer and a few comrades put their own lives at risk, saving multiple lives. For his efforts he was awarded the Congressional Medal of Honor. Under the Uniform Code of Military Justice (UCMJ) Meyer technically violated regulations by failure to obey his command. Yet, his conscience would not allow him to live with the death of others for the sake of obedience to a governing authority. In this instance, the awarding of the nation's highest medal is a confirmation of conscience despite disobedience.

The Apostles Peter and John were thrown in prison for preaching the message of Christ. When threatened with further punishment they reminded the crowds that their loyalty was to a higher authority of which the accusers were not aligned. The Pharisees who were adamant about the letter of a law were committing idolatry because they had shifted authority from Scripture to themselves. Peter and John were both eventually killed for their faith. Conscience comes from right relationship with the Word and we are called to "keep the faith."

Galatians 5:18 says, "But if you are led by the Spirit, you are not under the law." Faith implies relationship, religion parades membership. Paul guides the church of Corinth saying that following all the rules and doing incredible things will mean nothing without love. Love is always intentional. Wages are exchanged for services but gifts are freely given. God makes certain throughout Scripture that obedience is more important than sacrifice. "Sacrifice" in this context was an act of atonement for sin. The Israelites took on the idea that their efforts mattered more than their allegiance. Obligation is about earning just-ification through works or to avoid suffering. Obedience reflects the heart.

One of the grievances Martin Luther held in breaking away from the Catholic Church in the 16th century that led to the Protestant Reformation was the Church's use of indulgences. Indulgences are essentially a way of reducing suffering while in purgatory (the issues being a merit-based system and purgatory—both of which are biblically unsound). A bizarre abuse of this practice is observable in our modern age as Pope Francis offered indulgences for Twitter follows in 2013. The lowest form of obligation is acting out of fear of repercussion.

Jesus was challenged for breaking religious laws and associating with sinners. The Pharisees sought ways to convict him for violating the law of God yet were only able to convict him of violating their man-made rules. The law of grace we have entered into with the new covenant is about the posture of our hearts. The fruit we produce will reflect a relationship with

Christ if our hearts are aligned with his Word. We cannot reflect God's heart if we do not honor his commands. Relationships move us to action. Jesus gives us a metaphor to consider in Matthew 21:28-32:

> "'What do you think? A man had two sons. And he went to the first and said, 'Son, go and work in the vineyard today.' And he answered, 'I will not,' but afterward he changed his mind and went. And he went to the other son and said the same. And he answered, 'I go, sir,' but did not go. Which of the two did the will of his father?' They said, 'The first.' Jesus said to them, 'Truly, I say to you, the tax collectors and the prostitutes go into the kingdom of God before you. For John came to you in the way of righteousness, and you did not believe him, but the tax collectors and the prostitutes believed him. And even when you saw it, you did not afterward change your minds and believe him.'"

Words convey intention, but actions reveal the heart. Our love is sacred to God because it is the only gift we can bring without obligation. Love is always freely given, but it must be honorably given. Love is not the aimless outpouring of emotion—it is self-controlled. When I stepped into a personal relationship with Christ, I discovered a desire for God's Word. Consequently, my immediate response was somewhat polarized in that I carried a disdain for organized religion feeling that it was a barrier to relationship.

Various denominations of Christianity have practices that are relevant to their respective cultural contexts. I grew up in a denomination that held rules regarding dancing, movies, and alcohol for example. Dancing was prohibited because it led to sexual temptation. Going to the movies was a bad influence and financially supported sinful institutions. Alcohol was to be avoided because it could prevent one from having a sober mind. These rules were established to guide congregants away from sin but morphed into a rejection of culture. Dancing, watching entertainment, and consuming alcohol are not prohibited in Scripture but are cautioned against.

When extrabiblical practices persist in a church they essentially become the laws of that denomination. Following God can feel like an obligation when man-made rules are the focal point of relationship. For this reason, I am not a fan of denominational constraints. But I do have great respect for structures which seek to strengthen a relationship with Christ over a relationship with the world. This is where the non-denominational movement suffers from a lack of accountability. The emphasis on relevance at the expense of structure is the inverse side of the religious pendulum. Loving our neighbors without urging them toward repentance helps no one. *Why* we do what we do matters more than *what* we do. Without God's Word, we lose our *why*.

All Scripture points to the person of Jesus. The progressive movement seems to hold the belief that it is Jesus who helps us understand Scripture. The problem with this idea is that it takes the behavioral character of Christ as a lens in which to interpret the Bible. Not only does this strip him of his divinity, but it separates the character of God seen throughout the Old Testament from the loving and kind Jesus of the New Testament (formally known as the Marcionite heresy). The law was given to convict us of our sin and point to our need for a Savior. There is a transition from tradition to grace when that Savior comes and relationship has always been present on both sides of the covenants. God did not create us to transact with him but to be in relationship with him.

Old Covenant

The book of Exodus illustrates God's struggle in shaping the Israelites as his own children. The Israelites were enslaved for 400 years in Egypt under an oppressive system with its own cultural practices and pagan gods. God delivered them from oppression and led them into the wilderness. This large group of people wandering in the desert was a predominantly lawless community with no framework on how to live as free people. The Mosaic law was then given to the Israelites which displayed God's authority over his chosen people. The Exodus from Egypt moved them into a theocracy. In a theocracy, no separation exists between church and state as we know it today. Laws and commands come directly from God, not a governing body.

Intimacy with God sometimes means isolation from the world. In order to be righteous before God and enter into his promised land they needed to be cleansed from all other cultural influences. They also needed a behavioral framework. They weren't given a book of hundreds of laws immediately. The law began with the Ten commandments. These commandments addressed a concise moral standard for a generation of people who have never known freedom:

1) You shall have no other gods before me
2) You shall not make idols
3) You shall not take the name of the LORD your God in vain
4) Remember the Sabbath day, to keep it holy
5) Honor your father and mother
6) You shall not murder
7) You shall not commit adultery
8) You shall not steal

9) You shall not bear false witness against your neighbor

10) You shall not covet

The first four commandments illustrate behaviors toward God, while the remaining six illustrate our behavior toward others. Every commandment is relational. The Israelites were developing the character of God by following his commands. This was the initial framework for their interpersonal dynamics. Mosaic Law was designed to keep the Israelites separate from all other nations until the Gentile world could inherit the gift of salvation. The Israelites were given moral, ceremonial, and civil laws. Moral law developed the character of God. Civil law defined what was acceptable and harmonious. Ceremonial law defined the culture and foreshadowed Christ to come. Civil and ceremonial law went away in the new covenant while moral law remains.

Adherence to God's law was required for all generations leading to the fulfillment of God's promise. The Israelites did not understand God's intention to bless all nations. God displayed his power to the world through his people by requiring them to shed all other cultural practices and influences. The Israelite's victories and survival could not be attributed to any other source but God as they lived in accordance with his law. Yet, they undervalued God's commands because their hearts were infatuated with pagan influences. The Israelite's regressed into idolatry and worship of foreign gods. They observed practices that God specifically forbade. They deviated from the law for what appeased their sensibilities and honored the created rather than the Creator.

Renowned child psychologist Jean Piaget developed a theory of cognitive development that outlined the different stages of human intelligence. Through years of researching children, Piaget identified developmental stages of intellectual comprehension. The four stages he discovered are: Sensorimotor (Birth-2 years), Preoperational (2-6 years), Concrete Operational (7-11 years), and Formal Operational (12+ years). Similar to human development, our faith also progresses in stages.

Children in the Preoperational stage think symbolically before they can think cognitively. This means that children don't develop the capacity for rational thought until the Concrete Operational stage. In this stage, children's thoughts are ego-centric. They believe other people see, hear, and feel exactly the same as themselves. Adults enforce acceptable behavior in primitive terms, using rewards and punishments during the Preoperational stage.

Say you have a 4-year-old boy who is disputing with another child around the same age over a toy. Your child took the toy from the other child and is reluctant to share. You will likely have a hard time explaining to him that the root cause behind the dispute is his selfishness because he sees the world through egocentrism, believing that other children should feel happy when he is happy. He hasn't developed the capacity to see the world from any point of view except his own. The capacity for abstract thought and reason develops around age seven.

The most effective way to encourage behavior in the Preoperational stage is to appeal to egocentric thought. Saying something like, "You should share the toy because you wouldn't like if he took it from you" to the child will be ineffective because pretending to be in someone else's shoes is an abstract principle he cannot comprehend. It's common to tell a child in this scenario, "because I said so." They may not like it, but they understand authority because it is a burden to their will. Rewards and punishments are the language they understand. As a child matures into the concrete operational stage they develop the ability to reason and the capacity for abstract thought. The goal is to raise children to become relational adults who control their selfish impulses.

Psychology helps illustrate the innerworkings of humanity, though it does not interpret Scripture. It simply validates biblical truths in some areas, illuminating the complexity and mystery of our design. In the context of Exodus, the Israelites relationship with God as Father was essentially in the Preoperational stage. They did not value him because his will conflicted with their cultural desires. The language of the Ten Commandments is void of abstraction in order to be clear in expectation. God's directives are authoritative: "Because I am the Lord your God." The language they understood had to be primitive. They were to keep God's commands because he said so. If they did not keep his commands, then they might be put to death or be cut off from their tribes.

Strict religious structure and avoidance of all external cultural influence was essential in Israel's theocracy. The Old Testament is clear on why the Israelites were restricted from external cultural influences. Culture appeals to us more than religion because humans are experiential before we are relational. The Israelites could not isolate the corrupting influence of the pagan world without a structure. Israel lost its theocracy with God when they failed to recognize him as their ultimate authority. The theocracy was designed to last until Christ's death on the cross, which fulfilled the Mosaic law and the promise God made with Abraham to bless all nations.

New Covenant

The original 613 laws given to Moses included civil,

ceremonial, and moral instruction. The Ten Commandments are 10 baseline moral laws. Mosaic law was established roughly 1500 years before Christ and foreshadowed his coming. During Jesus' life he was asked which of all the laws was the greatest. His response is seen in Matthew 22:37-40 where he reduces all of the laws of Moses down to two categorical laws:

1) You shall love the Lord your God with all your heart and with all your soul and with all your mind.
2) You shall love your neighbor as yourself.

Jesus says the entirety of the law is dependent on these two commands. Because moral law is forever binding, the Ten Commandments are summed up with these two directives–the first relating to God, the second relating to others. The Mosaic law established with Moses existed until Christ established the new covenant which would spread to the Gentile world. The new covenant marks the beginning of Christianity and the end of Judaism.

After the abolishment of Jewish civil and ceremonial law, a need existed for a replacement civil structure. In the transition from Judaism to Christianity, Jesus reaffirms which commandments would augment secular governments that are intended to be the replacement. The struggle between Christ and culture is made evident in Romans 13 as Paul demonstrates the overlap between secular government and the Christian faith. Israel lost their theocracy with God when they demanded a king. They suffered under kings who did not observe God's commands, were oppressed by other nations, and ultimately lost their independence to the Romans in 63 B.C. The Roman governance over the Jewish people paved the way for God's new covenant and was the bridge connecting Christianity with culture.

There is a significant power shift that takes place in the

years surrounding Jesus' life when the Roman Empire and Judaism converge in Israel. The Roman Empire legalized Judaism as a religion in 27 B.C., but also painted Caesar as an authority over the Jews. They were allowed to engage in religious worship so long as they were peaceful. This intersection between religion and government represents a separation of church and state, a conflict between Christ and culture. Ironically, the Jews rejected secular governmental authority even though they lost their theocracy by demanding a human authority. From the 60-year period between 30 B.C. and 30 A.D., Jewish ceremonial and civil law coexisted with Roman law. The fulfillment of the Mosaic law at the death of Christ marks the integration of the Jewish and Gentile peoples into the Christian faith.

There is an interesting moment when the Pharisees question Jesus about who holds more authority, God or Caesar, asking if they should pay the Roman imperial tax. Jesus' response is unique in recognizing who the ultimate authority is. Jesus tells them to bring him a denarius and says, "Whose image is this? And whose inscription?" If humanity is created in God's image, then so is Caesar, signifying that God is the higher authority. It is a relevant question considering the Jews historically operated within a theocracy and believed they were the only chosen people of God. Jesus clarifies God's authority above Caesar's though they are still subject to the secular governing authorities.

Jesus spent his life modeling the Jewish law without ever violating it. In fact, he fulfills the law, never violating secular Roman law, and brings humanity under the law of grace. While in custody of the Roman government Luke 23:4 displays Jesus' civil innocence as Pilate found no grounds for charging him. He was being convicted of violating man-made religious rules. The religious elite manipulated religious law for their own benefit and condemned others with legalistic expectations. The Pharisees were blind to the Son of God because they became the gate keepers of salvation.

Jesus was considered a heretic because he claimed to be the only way to be justified before God. Their competition with God for authority was nothing more than idolatry. Jesus reveals himself to us through the Spirit, and his law is written on our hearts. The Spirit is who guides our conscience and leads us to God's authority in Scripture. We are not bound to the ceremonial and civil laws of religion. We live first by the authoritative Word of God and secondly within the civil laws of secular government. Christianity and culture are intended to be complimentary. Should conflicting issues arise, it is our duty to remain obedient to the Word of God in Scripture as the ultimate authority.

CHRISTIANITY AND USMC CULTURE

Nothing which implies contradiction falls under the omnipotence of God.

– St. Thomas Aquinas

Church is more than a social club on Sundays. Christ distinguishes us in all circles and guides our behavior in various cultures. I am grateful that I was raised with a Christian found-ation, even if it was a bit traditional. I found myself confronted with a cultural paradigm after high school. The Marine Corps had its own unique culture which was more rigid than any religious denomination. I was expected to adhere to all rules and regulations in addition to my own Christian faith. Attempting to obey the regulations of both led me to a deeper understanding of what it means to be a Christian under secular rule.

The military takes people through a myriad of cultural dimensions. Much like organized religion with beliefs, rituals, and structure. It is a formal institution with its own judicial system, traditions, and purpose. Integrating the military with Christianity blended two significant authorities. Some believe these authorities are at odds with each other. The beauty of Christianity is that it is a system of moral law which has been integrated with civil government. In the church age after Christ's death, secular government replaces the civil law portion of Judaism. Biblically speaking, nothing prevents a Christian from serving in the military. The two exist together though separate, much like the church is separate from the state, yet not adversarial.

Deployments add another dimension to this integration—cultural complexity. Obeying God's Word, serving in the military, and living in a foreign country put a strain on the practice of faith. This is a great example of why Jesus simplifies the commandments down to two all-encompassing principles. Our time in the field and conducting missions overseas made it difficult to maintain a consistent rhythm to practice faith. Most adapted without conflict, but those tied closely to religious practices carried a heavier burden in balancing both.

Chaplains were an available resource for spiritual development and conducted services at Hurricane Point in Ramadi. However, I found myself rarely going. I spent more time reading my own bible, listening to music, and engaging in conversation around the firepit with other Marines. I grew closer to God in the company of other Marines than I did seeking my own spiritual development with the Chaplain. The Chaplain was a great theological resource, yet sharing what little I knew with others filled the void of a formal congregation. It was in my relationships with others that I became the body of Christ.

I was further along than others in my faith journey because I had grown up in church. The Chaplains were able to help

188

Marines who were spiritually struggling with deep questions that I could not answer. Our exposure to violence stripped us of the common inquiries such as how we should tithe or what the bible says about dating. The hostility of Iraq reduced us to a primitive faith where we wrestled with the existential realities of life and death: *Is there a God? Is there life after death? Can Christians kill?* I noticed how the issue of violence inevitably measured itself to the moral law of God. Then I understood the connection between military service and faith clearly.

I rarely wrestled with having to take life in the course of my duties. In fact, most Marines had no reservations on pulling the trigger. Regardless of the level of faith each Marine had, we all sought answers in the metaphysical realm. Working through psychological and emotional rationalizations proved to be a bigger challenge than theological justifications. Spiritual connection was black and white once a life had been taken. Marines either ran toward God or away from God. We are all built with an *intrinsic* repulsion toward killing, even if it is morally justified. To kill is not necessarily sinful, but we still bear an emotional burden because all humanity bears the image of God.

The beliefs we have about God determine how we navigate the world and all its chaos. We all sought relief in a higher power during moments which we thought would be our last or while witnessing the last moments of someone else's life. When danger was present we were focused on survival. The threat of death forced us to accept that we are mortal, so we turned to the miraculous for hope in dire circumstances. The intimacy of death brought us to an acute awareness of our need for God. In our moments of greatest desperation, we seek a relational God who can restore us. These unique experiences are not exclusive to military culture. What we believe about God is revealed when we are in duress.

James 1:27 shows us that true religion is relationship. Religion and relationship are not mutually exclusive. Christianity emphasizes obedience through relationship. We are called to serve others; not to gain salvation, but to reveal the source of our love. Rules and practices are designed to lead us toward proper relationship. The ultimate display of God's love for us is when his mercy triumphed through judgment at the cross. If legalism is the only way, then mercy is breaking the rules. But if mercy is all we preach, then justice becomes impossible. The conflict between religion and relationship is satiated when our consciences are informed by the Word.

CHAPTER THIRTEEN
THE UNCONVENTIONAL

Sometimes a scream is better than a thesis
  – Ralph Waldo Emerson

WHEN I WAS IN middle school, my mother would play music on the radio while driving my brothers and I to school. There were roughly two stations for each genre on the FM channels. I developed a preference for Rock in junior high. That's when the conflict began. Kenny G wasn't the talk around school. The friends I hung out with listened to the local Rock station in Los Angeles. I suggested this preference in the car one day. My mother switched to the local rock station but changed it back because it was "too heavy" and gave her a headache. When I persisted, it suddenly became a moral issue. She explained that it was not a good influence and that I shouldn't be listening to songs with swearing and promiscuous content. But how would she know that unless she listened to it?

I inevitably listened to rock bands all throughout high school. By my senior year I learned to play my favorite songs on guitar. When I discovered Christian rock bands, I felt like I had stumbled upon a solution to my mother's moral dilemma. Hardcore music hit the scene, and I ventured into a new sound that was sure to upset her. I played a few songs, and she instinctively jumped to the bad influence argument adding, "you can't understand what they're saying!" I read a few lines of the lyrics to reveal the biblical principles, but she defaulted to the headache response. I was finally free to pursue my own musical interests and broke out of a generational barrier.

BEYOND EXPECTATIONS

For as the heavens are higher than the earth, so are my ways higher than your ways and my thoughts than your thoughts.

– Isaiah 55:9

Music is a unique human expression that sounds different in every culture. It takes us back to places that we wish we could live permanently. It can also take us to places we never wish to return. What shapes our music preferences? I think we gravitate toward music that we associate with the happiest times of our lives. As communities and technologies evolve, so do the sounds we admire. The expression of worship through music naturally reflects the culture by which it is embedded.

I grew up in a church where worship consisted of a choir and piano player. Every Sunday we would gather together and open a hymnal with hymns written anywhere from fifty to a few hundred years ago. Amazingly, many hymns are still in circulation today. They have even been remixed in a modern format known as Contemporary Christian Music (CCM). The lyrical content of these old hymns persisted despite numerous cultural shifts. What I find particularly interesting is how long many hymns have been sung in an outdated composition through the early 2000's. As CCM grew in the 90's and early

2000's, a massive cultural clash bemoaned the emerging cultural expression.

Full bands invaded the sanctuary and drove the older congregants into a hysteria. The uprooting of a preferred style of music was not well received–it was even seen as heretical. Generational transitions sometimes involve breaking the barriers of preferred expressions of faith. Worship through music is just one area of observable shifts within church culture. The future of the church is always found in the youth it cultivates. But what happens when a preferred expression is being phased out? There are always going to be generational shifts that bring about different expressions of faith. The way we deal with adversity is shaped by how we relate to others. Relationships define culture.

Take aggression for instance. I struggled with aggression outside the military even though it was normative and encouraged when I was actively serving. The key indicator was the culture. The ability to release frustration without judgement in the military helped me process intense emotions in an accepted way. Outward frustration was not only common, it was communal. It created stronger bonds through shared trials. Civilian culture has a low threshold for this particular emotion. Outside the military, venting through frustration is not a socially shared catharsis.

My expectations outside the military were gravely misaligned. Without the ability to release frustration, I could explode unexpectedly or descend into apathy. I've damaged a few relationships from unrestrained frustration. While anger is almost wholly rejected in civilian life, it was a common response to the problems we encountered on deployment. Anger is a cry for help when our rational faculties are compromised. Unlike apathy, anger is a sign that someone still cares. Frustration means someone is still in the fight though others may not know how to receive it. In relation to hardcore music, listening to prayers screamed over distorted guitar and double bass pedal changed my perspective of constructively channeled aggression. Music was a gift when stress overwhelmed me.

Music brought me closer to God during the military. The isolation in Iraq away from family, friends, church, and a spiritual leader left me on my own to figure out how to further my relationship with God. I was hungry to know God and how to navigate the chaos around me. I knew the military was where I was supposed to be in that season regardless of my doubts. I turned to music as a way of escape. I loved the sound of indie guitar riffs and irregular drum beats. Uncommon chord transitions and heavy distortion eased tensions inside me when words had no effect. Listening to other musicians express emotion in an artistic way was a relief. The heavy dissonance of distortion and screaming brought peace. Lyrics that layered the music brought everything together with a strange comfortability.

The intensity and aggression of screaming and distortion is unrelatable to most people. This expression of faith caters to a small crowd. It seemed that only grungy outcasts and disgruntled teens were into this emerging genre of music. The older generations proclaimed that there was nothing "pleasing to God" about it. Their rejection of it made it more appealing to me. I was fascinated with the passionate expressions of anger as a way of crying out to God. Not all the music I listened to was heavy, but it was predominantly hardcore music that opened the door to a world that significantly shaped my faith when I was isolated from normalcy.

Most people are disturbed by aggression in music. It's definitely not for everyone. It wasn't something I just liked one day either. It took a few years of listening to indie-label bands to end up down the rabbit hole of metal, screamo, post-hardcore and whatever other genre designators the music industry created. My youth leader in high school gave me my first hardcore album a few months before I left for boot camp. It was loud, angry, dissonant, but most of all—unconventional. It became a prominent influence when my faith was tested.

NON-TRADITIONAL FAITH

The most painful state of being is remembering the future, particularly the one you'll never have.

— Søren Kierkegaard

I was in a dark place when I shipped off for my first deployment to Iraq. I had just experienced my first break-up with a gal who became engaged to someone else in the latter phase of our relationship. In the midst of saying goodbye to family and friends, I had no time to emotionally process a shattering heartbreak, much less the reality that I may not make it home. My anxiety was amplified the first week we arrived to Ramadi. We suffered our first casualty hours before my first patrol, and an IED that nearly destroyed the vehicle in front of me. The anger and fear I felt in that moment was like nothing I've ever experienced even to this day.

It was as if time stood still after that patrol. I needed an outlet and music was the only way to reduce a hemorrhage of emotions. I vividly remember sitting on my rack with head-phones on to drown out the world. I was infuriated and unable to purge a dissenting rage. I had no shelter to run to. The loss of Swanberg, the last memories of home, the rejection of intimacy, and no promise of a future brought me to a state of complete apathy. I felt hollow and helpless to change anything. The existential hopelessness of that moment suppressed the fear of death and all the aggression I had left. I could have been on fire and yet would have gone out quietly. In the silence of that moment, it felt as if God was speaking through the lyrics of a shuffled playlist.

The words "just enjoy it" resonated in my soul from a specific song. I had listened to this song hundreds of times before, but this time it clicked. I was led to Ecclesiastes 8:15 and reflected on where God had been. A thought from a few hours earlier came to mind. After they told us that we had taken our first casualty, I sat in the driver seat before that patrol and prayed while waiting for clearance to depart. The fear of Ramadi infiltrated my mind. I was nervous and excited. The unknown of all unknowns sat right before me. Life and death. A strange peace came over me as the diesel engines of our Humvees idled. I somehow felt present.

All anticipation evaporated at the sight of Iraqis going about their lives. Something changed inside me when a little girl poked her head out and looked at me while staged behind the government center. She had no fear of us and smiled as if the world around her wasn't tearing itself apart. We were the ones to be feared, yet here I was with an internal world as dark as her external one. Somehow, it didn't steal her joy. She waved then ran back into her house.

God is always fighting for our attention. One song was enough to ease my mind and remind me of my purpose. The difficulty lies in being stripped of everything to willingly surrender before Him. The apathetic state I descended into rendered me helpless and on the verge of hopelessness. The only thing that kept me from becoming destructive was the nearness of God through music. I had every reason to be angry and lash out. I had more than enough pain to walk away from God and give up on my faith. I couldn't comprehend the purpose of what I was going through, but I still believed God was doing something, even if it would be a long time before I could see it. If faith was all I had left, then I wasn't far from confirmation or a grand illusion.

I bought a used 1ˢᵗ generation iPod from a friend right before our first deployment. I loaded every album I had onto it. Whenever I needed to escape from reality I put my headphones on and drifted off to the sound. Our unit was attacked every week in country, but I found myself fighting more spiritual enemies than physical ones. I listened to so much music that I could quote a line from a song for nearly every problem in life. Lyrics offered tremendous wisdom. The faulty brick of an iPod that held hundreds of albums became my sanctuary.

Many of my beliefs about love, relationships, faith, and the world were shaped through the music I listened to in Ramadi. Later, they were either validated or refined by Scripture. Some of the music was worship but most was an array of artists from soft jazz to heavy metal. Not every artist was explicitly Christian nor every lyric spiritually informed. Sometimes just the sound of music was enough to bring peace. There were a number of artists who were Christian but not Christian bands. By exploring the lyrical content, I began to discover biblical origins in quite a few songs. One heavy song in particular was a direct reference to Matthew 8:22.

In this verse Jesus is approached by people who say they will follow him but are distracted by other tasks. Verse 22 says, "But Jesus told him, 'Follow Me, and let the dead bury their own dead.'" The remainder of the song is a creative dialogue of what God is communicating to those who choose to follow him: Are we in it for the harvest and not the famine? Are we willing to give up what we love most should God require that of us? The cost of following Jesus comes with a price, yet there is still his promise that we will never go beyond his hand. And whatever we do, he loves us despite our darkest moments.

I didn't have the liberty to go to a church on Sunday or meet with a small group as I did in high school. Yet I felt the presence of God when I listened to music or played guitar. I discovered the depth of his love in isolation from the world. Music was an effective tool in my spiritual development. The conflict in Ramadi couldn't separate me from God. I developed a burning desire to know God because he was all I had. No family or friends, significant other, pastor or youth leader. I had a direct line of communication because I was listening. I had everything I needed to fight the spiritual battles I was in. I was even equipped to help others fight their battles.

As Christians we spend our lives drawing near to God. Music is one of the most effective ways to do so. It doesn't matter if the music is the screeching destruction of our vocal chords or the whisper of an apathetic breath. God is moved when our hearts appeal to him with praise. God cares about the posture of our hearts and is glorified by the creative ways we express our love for him. God is moved even if we have no musical talent. Simply listening to music can help us realize his eternal presence. Music is a language of the soul because it is a weapon in the spiritual world. As David used the sword on his physical enemies and the lyre on spiritual ones, we also are equipped with weapons for both worlds. Music aligns our spirit with God's so that we can discern how to effectively engage in the physical battle.

Music is a weapon of spiritual warfare. It would be suicidal to have gone out on patrol with my guitar preaching love and peace. As skilled as David was as a musician it was a well-placed stone that defeated Goliath. The Apostle Paul advises us that though we are spiritual beings, we dwell in physical bodies. We all must uphold the responsibility of physical defense. Levels of physical proficiency vary greatly, and most are not called to the formal vocation of defense. Those who are called carry the burden of violence and all that it encompasses. We hope for peace while we train for war because one day the enemy will show up. However, we must prepare to fight the spiritual battles first before we engage in earthly ones because the primary war we fight is spiritual.

We are made in God's image and radiate his glory whether we realize it or not. Proof of God is all around us. Yet, the way we determine God's *truth* is by testing everything to Scripture because it is God's authoritative Word. The lyrical content of the music I listened to reflected what I read in my Bible. I didn't know where the singers were coming from when they wrote these songs, but they were speaking things I needed to hear in moments I needed to hear them. They often led me back into the Word.

When we find ourselves utterly powerless, where nothing can satiate the pain we are feeling, our hope can be found in God's Word. The authority of his Word has the power to move us out of the agony of our worst suffering. It is the only thing that can heal us when our suffering transcends the reality of everything we have ever known. Additionally, music holds power in the physical realm that influences the spiritual world. It's a way to honor God despite our physical limitations. Whether we have the ability to play an instrument, write lyrics, sing, or simply listen to music others have created, we are engaging with powers in the spiritual realm. Instruments and vocal chords become weapons to be used for God's glory.

WORSHIP AS A WEAPON

Yet You are holy, enthroned on the praises of Israel.

— Psalm 22:3

God inhabits the praises of his people. God is moved by our worship. Even before we were created God enjoyed the sound of music. Job 38:6-7 says God played his heavenly Spotify shuffle while he created the world; "On what were its bases sunk, or who laid its cornerstone, when the morning stars sang together and all the sons of God shouted for joy?" How much more pleased must God be to receive praise from our own free will? God can command the angels and the heavens to make a joyful sound, but it is our willful praise that captures his attention like nothing else. He draws near to us when we draw near to him.

In II Kings 3:15 Elisha the prophet requests a musician play for him in order to receive a prophetic word from the Lord. In this chapter, Elisha has been approached by three kings for a word from God. Elisha was annoyed with them and initially directed them to find an answer from someone else. Due to his respect for King Jehoshaphat, he proceeded to complete their request but not before calming his spirit to the sound of music. It can be assumed that his frustration with the kings caused his mind to be distracted. He called upon a musician to bring his spirit back into balance in order to hear from God. Elisha knew the presence of the Lord could be found through music.

Music also has the power to unite people and open doors. 1 Samuel 16:14-23 recounts when King Saul lost God's Spirit and underwent torment by an evil spirit. In response to this torment, Saul's servants insisted that he have someone play the lyre for him. Even Saul's servants understood that music was the way to engage in spiritual warfare. Saul listened and David was brought in to play the lyre. The spirit would leave Saul every time David played for him. Through David's musicianship, Saul was blessed and able to be near the presence of the Lord again.

It was David's musicianship that also positioned him for his future. He was brought in to play the lyre and quickly became Saul's armor bearer. His presence in the King's court led him to confront the problem of Goliath. After defeating Goliath, David was made a chief in Saul's army. David's proficiency with an instrument opened the door for his proficiency with weapons to be realized. David's time in the wilderness strengthened him in spiritual warfare before physical warfare. Then he fought Israel's enemies with swords and a sling. When Saul experienced torment by an evil spirit David used an instrument to drive it away. Saul eventually drifted away from God's spirit that he tried to kill David, turning a spiritual threat into a physical one. Though David was more skilled in physical warfare than Saul, David would not harm him because he respected the Lord's anointed one. David had no choice but to remove himself from the situation.

We don't always have the opportunity to remove ourselves from the situations we are in. Being deployed to Iraq was one of those times. Physical and spiritual battles were fought simultaneously when there was no escape. When I played guitar around the firepit in Ramadi, Marines would sit in reverent silence. I often played contemporary worship songs I learned in high school or songs a fellow Marine and I wrote. Strumming through steal strings while sitting around a fire pit with gunfire and explosions in the distance held us in spiritual balance. The presence of God was a tangible experience that transcended violence. Regardless of what faith was present in the circle, we we're all reunited with the Creator.

CHAPTER FOURTEEN
GOD AND VIOLENCE

So the LORD said, "I will blot out man whom I have created from the face of the land, man and animals and creeping things and birds of the heavens, for I am sorry that I have made them."

– Genesis 6:7

DESMOND DOSS WAS AN AMERICAN WWII combat medic who received the Medal of Honor for his actions during the war. He was also an extreme pacifist. He refused to carry a rifle despite choosing to serve at the frontlines of battle. Doss is highly honored for his bravery to serve despite his moral opposition toward violence. While his service is exemplary, a battalion of pacifists would have lost the war. Some Christians believe the taking of life is never justified. To them, the difference between murder and killing is indistinguishable. Some understand the role of a nation's military but choose not to be a participant. There is nothing wrong with this understanding so long as others are willing to fill the ranks.

I often hear biblical references supporting pacifism. Yet, I am not convinced that pacifism is appropriate for all Christians. Living a vow of non-violence is fitting for some people, but it is not a sustainable solution for all. God chooses to complete his work in the world through us. He has the power to heal all sickness and disease, yet enables us to become doctors and create medicine. He has the power to sustain life in every corner of the world yet empowers us to become humanitarians. He has the power to end all wars yet builds us up as warriors. We serve a God who is righteous and just. The God of the Bible is not all love and no wrath. The question is who will enact his love and who will carry out his wrath?

The authority to take life is a great responsibility that God has authorized us to use for his justice. I'm thankful to have grown up with a father who understands the judicious use of force and willingness to use it. I remember having to sleep in the living room for a few days when I was 7 years old, barricaded behind the couch next to my father, brother, and pregnant mother. I vaguely remember the odd change in routine and never understood why until I brought up the incident with my father. The memory came up in a conversation where he explained that someone made threats to kill our family. This is how he described the incident:

> I knew a friend in high school; he and I were involved in choir, though he was a year younger than me. After graduation, I didn't see him until 14 years later. He arrived at church one Sunday with a wife and two children. He was wearing dark sunglasses, had tattoos, and my initial impression was that he just got out of prison. I greeted him and he came to the Sunday school class I taught. He became heavily involved in drugs after high school and I got the impression that he had mental health problems. I noticed in conversation that his theology was warped and manipulative.
>
> He also told me he had done 10 years in prison for

manslaughter. He was sharing an apartment with a room-mate and believed his roommate ripped him off by stealing his drugs, which led him to subdue and torture him to death. During his last year in prison, he realized he would be 30 years old and that he had wasted his life up to that point. So, he committed his life to Christ and spent most of his free time with the Chaplain. He also met his wife through a friend while in prison. Then, fresh out of prison with a wife and two stepchildren, he decided to come to church where I ran into him after 14 years.

One Sunday afternoon his wife called the church. She said that her husband hadn't changed, was using cocaine, and sent her out without the kids to shoplift merchandise and return it for cash. She was told to "not come back until she has $200 in her pocket." She instead drove to the church for help, where the pastor called me. We hurriedly helped her and the kids gather some things and moved them into a women's shelter. Then, we arranged to relocate her and the kids out of state with some friends who arranged housing and employment for her.

After roughly three weeks of not being able to locate them, he called me while intoxicated. He explained, "I need to be back with her by the end of the month or I'm coming after your family." He said he had a shotgun, a handgun, and a different car so I wouldn't know it's him if he arrived. He made it clear that he would take my wife and kids because I took his. I filed a report and the following day and the police had a $2 million warrant for his arrest. In the meantime, I contacted your school and provided this information with instructions to immediately call 911 if he showed up. I started making plans to relocate you, your brother, and mom out of the area until his arrest.

For the next two nights I slept on the sofa in the living room with a handgun in case he showed up unannounced. I would have shot him if he showed up. A SWAT team was able to arrest him, where they took him to the county jail and allowed him a phone call. He called me at home acting friendly again. He said he was just kidding and wasn't serious about the threats. He was sentenced to 1 year in prison with a notation on the case for me to be contacted pending his release. I was contacted by state parole after his sentence who told me, "He is being released, but we will place surveillance on him."

He was scheduled to go into a 90-day lock down facility for drug rehabilitation. However, he was removed from the rehabilitation center after he violated their terms of service. His wife informed me two weeks later that he was dead. He ended up in Los Angeles' skid-row hotels with prostitutes using drugs again. After assaulting one of the prostitutes, she called the police to report the abuse and, on their arrival, explained to them that he "would not go back to prison." The officers entered the hotel room and ordered him out of the bathroom where he was sitting. He refused, and slashed at the first officer with a knife. The officer quickly fired two rounds striking him in the chest and killing him.

The threat of violence didn't come from a distant enemy or a random stranger set out to burn the world in his rage. This was a former friend. Evil stems from the heart of corrupted men and can reach a point of no return. I wouldn't be here without those who understand and combat violence. Thankfully, law enforcement prevented further tragedy. Without them, my father was the only defense my family had against a murderous felon. Judicious violence eliminated the danger after other forms of correction failed.

PATHWAY TO VIOLENCE

Stars, hide your fires; Let not light see my black and deep desires.

– Shakespeare

The pathway to violence is a treacherous road which begins with small sins. These small sins progress into unthinkable horrors. All these sins are birthed from a lack of empathy. It is the erosion of empathy that leads to a complete lack of remorse. Violence is not the problem we face; a lack of empathy is. The level of someone's empathy and remorse indicates the severity of their violent potential.

The notorious serial killer Ed Kemper exemplifies the horror produced by a lack of empathy. Born in Southern California in 1948, Kemper faced a barrage of emasculating assaults from his mother, which began a violent grievance. He mutilated his sister's dolls as a child, buried a cat alive at ten-years-old, and murdered his grandparents at fifteen. He exhibited "rehearsal" behaviors, effectively building up the courage to commit worse acts. His emasculation and abuse came from the feminine, which became the target of his aggression and the source of his deviancy. From inanimate feminine objects, to animals, to people, he acted violent towards all things female. By twenty-four years old Kemper's progression of violence escalated to the gruesome mutilation of eight female college students, his mother, and her best friend.

In a 1991 interview, Kemper talked about the murder of his mother. During the interview, a slight emotional slip can be seen as he describes his decision to kill her. Up to that point he had taken out all his aggression on feminine representations of her. What he really wanted was a mother who loved him, who saw potential in him, who encouraged him. Instead, he weaponized his wounds, producing the deepest human darkness possible. But by taking the life of his mother he removed any possibility of ever receiving what he craved. That is the full consummation of violence in the human heart. It doesn't take a lifetime to reach the point of no return.

Another example of a young offender is seventeen-year-old TJ Lane who walked into a school cafeteria in Chardon, Ohio in 2012. He shot four students, killing three and paralyzing one. Authorities believed that he killed for notoriety and exhibited warning signs before the attack. For example, Lane made a post on Facebook saying, "Die, all of you," and had a history of juvenile violence as well as an unstable home life. He descended the path of no return before his eighteenth birthday. At his sentencing, Lane laughed at the victims' families and said, "This hand that killed your sons now masturbates to the memory. F--- you all."

The warning signs of someone going down the path of no return are evident if we pay attention. Conduct disorder at a young age is a precursor to anti-social behavior as an adult. Both Kemper and Lane killed as teenagers and would continue to kill given the chance. Violent offenders often show a history of deviant behavior in adolescence. A common thread among murderers is the social environment they grow up in. They are not responsible for their unfortunate upbringing, but they remain accountable for how they respond. Sin is a conscious choice and violence is the full maturation of sin. Humanity does not lack remorse by design but rather is degraded by sin. God is adamant about fleeing from sin for a reason.

Adam and Eve's disobedience in the garden led to the downfall of mankind. Violence inevitably resulted as humanity drifted from God. In the story of Cain and Able, Cain became complacent with his sacrifice by not bringing his best before God. His laziness exposed him to the snare of the enemy. In Genesis 4, God confronts Cain about his disobedience. Verse 7 says, "If you do well, will you not be accepted? And if you do not do well, sin is crouching at the door. Its desire is contrary to you, but you must rule over it." God encourages Cain to overcome the temptation to sin. Cain must choose between God's will and his own will.

The first punishment for man's disobedience was to work the ground in order to yield food. Cain's punishment for disobedience was that the ground will never yield food for him again. Cain's growing disobedience resulted in the death of his brother. Therefore, God increased the consequence for his disobedience. He chose to spare Cain but marked him as a wanderer across the earth. Cain's lineage lasts six generations until "instruments of bronze and iron" (Genesis 4:22) emerge. Some historians speculate that Cain's offspring developed the first weapons of war. The birth of sin led to murder within the first generation of humanity, and the capacity for warfare by the sixth generation.

It took 1500 years for humanity to descend into complete chaos before God flooded the earth. The Flood is the first account where God destroys his creation. He annihilated the earth with the exception of Noah and his family. God allowed time for generations of people to turn from their evil before it consumed them entirely. While God did not tolerate evil, his patience before the Flood was a chance for people to repent before his wrath could consume them. To murder another human is to enforce one's own personal justice and is an idolatrous attempt to dethrone God. Murder is the defamation of God's image and character. The victim bears the marred image while the perpetrator defames God's character. The taking of life was an authority exclusively reserved for God prior to the Flood.

God understands our proclivity for violence yet did not remove our capacity for it. After the Flood, he empowers us to enforce his wrath. Genesis 9:6 says, "Whoever sheds the blood of man, by man shall his blood be shed, for God made man in His own image." This is the first place in Scripture where God gives man the authority to kill under specific conditions. God required the life of any man who shed another man's blood. Biblically, we have become agents of his justice. Prior to the Flood, God alone held this authority.

The rainbow is a marker that God will never flood the earth again. He left us with a symbol of peace in our progression toward reconciliation. The Flood is ultimately a story of redemption. Since humanity became irreversibly corrupt, allowing them continual existence would have been reckless. Ending the permanent suffering that humanity created and redeeming Noah and his family are proof of a gracious and merciful God. He restored balance to the world and promises never to flood the world again.

Noah's family eventually populated the earth where humanity lived under one language and vocabulary. Humanity's corruption climaxed again at the Tower of Babel where mankind came together to idolize themselves above God. Genesis 11:4 says, "Then they said, 'Come, let us build ourselves a city and a tower with its top in the heavens, and let us make a name for ourselves, lest we be dispersed over the face of the whole earth.'" God's response highlights our tremendous power: "And the LORD said, 'Behold, they are one people, and they have all one language, and this is only the beginning of what they will do. And nothing that they propose to do will now be impossible for them.'" (v.6). God does not allow great power to be separated from his authority.

God's response is incredible and demonstrates the love he has for mankind. Still, he did not take their power away. Rather, he dispersed it. Humanity was never meant to achieve greatness apart from God. We truly are wonderfully, yet *fearfully* made. Our eternal design was fractured when sin entered the world. God used patient yet swift justice to deal with our sin. He will not allow sin to have complete dominion over the world through us. In reconciling us back to himself, God slowly extends authority to those who move in obedience.

The idolatry displayed at the Tower of Babel resulted in confusion and conflict. God was dispersive rather than violent. Roughly 185 years later, God judged the inhabitants of Sodom and Gomorrah for their sexual perversions. This act of justice was regional in comparison to the global punishment of the Flood. A unique difference in this story was Lot's protest against God's decision to destroy the cities. In Genesis 19:21 an angel responded to Lot's request, "He said to him, 'Behold, I grant you this favor also, that I will not overthrow the city of which you have spoken.'" This is the first instance in which God withholds his wrath in response to a human request. The town of Zoar was spared from God's wrath at the request of the created.

As violence erodes the world, those with God's character rebuild it. Lot could have saved his own life and remained indifferent towards others. The first instance of God's wrath after the Flood was met with opposition. Lot knew God was justified in his actions, yet God relented when Lot appealed to his mercy. Humanity gained the authority to take life after the

Flood. Lot displayed the will to preserve it. The consequence of sin is death. No one is righteous before God without Christ. God will ultimately destroy evil. We have received grace to live in eternal peace with God if we turn from our sin. At any moment God can end our lives and be justified in doing so. The more God's character is revealed through humanity, the more he entrusts his authority to those who bear his name.

God's use of violence serves an explicit purpose. Humanity's uninhibited evil desires brought such destruction into the world that they became unrecoverable. These stories convey that by rejecting God, humans can achieve corruption beyond correction. God's solution was to remove them completely. His wrath is unmistakable throughout the Old Testament. The earth was flooded as an act of justice, making God the first and only proprietor of a "scorched earth" policy. Despite the seeming contradiction of character, God's wrath against humanity is not incompatible with his love for us.

God would be justified in erasing humanity off the earth for breaking his commands. Genesis records that the punishment for disobedience in the Garden of Eden is death. However, God withheld his wrath and did not kill Adam and Eve. God postponed their eternal separation from him out of grace, allowing humanity to live despite their disobedience. The serpent was cursed, and the man and woman received punishments rather than immediate death. Suffering can be a blessing in contrast to death. The Apostle Paul says God is rich in kindness by showing patience, and withholding his wrath. When we avoid the topic of violence, we are psychologically distancing ourselves from the traumatic effects of death and suffering. But we are also weakening our understanding of justice.

We live with the assurance of safety in the United States.

Our concentric layers of protection are not consciously thought of until a threat arrives. And even when a threat does arrive, we hardly know about them. We have become naïve to violence because we've abdicated our responsibility to combat it to an elect few. Our military and law enforcement agencies represent only a small fraction of the country's population. Whether we find them just or not is a different matter. The fact is, we need them. They are the men and women who recognize the wolves waiting to devour us in our complacency. Not everyone is called to serve in a formal industry of arms to combat threats. But everyone who bears the image of God is in a war whether they know it or not.

The Bible is full of violent stories of unspeakable suffering. Violence however, is not the focal point of these biblical narratives–God's holiness is. The spiritual enemy's goal is to separate us from God. The further we move from God the more violent we become. Our abuse of power is the result of disobedience to God's moral law. Severity of violence is directly correlated with distance from God. In the likeness of God, we have been entrusted with a portion of his power. By developing the character of God, we begin to understand the context and authority for the justified use of deadly force.

The enemy wants us dead. He has been at war with mankind since the beginning. He is spiritual in nature but doesn't act supernaturally on his own. In fact, the only people the devil himself is credited with killing are Job's family members. The hardest reality for some to accept is that God has a higher body count than the devil and all humanity combined. Through lies and deceit the spiritual enemy influences the hearts of men and turns us against each other. Understanding the pathway to violence is our first step in defending the kingdom of heaven here on earth. Spiritual battles always precede physical battles.

DEVELOPING A WARRIOR NATION

The LORD utters his voice before his army, for his camp is exceedingly great; he who executes his word is powerful. For the day of the LORD is great and very awesome; who can endure it?

— Joel 2:11

Violence was first authorized after the Flood for a limited and specific purpose. It was strictly defensive in nature–to preserve life. The expansion of God's authority to kill occurred again after Israel's Exodus from Egypt. The Israelites carried out God's justice through violent conquest to receive his promises and to enter the Promised Land. From Noah to Moses, killing another human was only for sustainability (defensive). The role of killing after the Exodus expanded to offensive violence. After God established the law with Moses, God began to make warriors out of the Israelites.

God heard the cry of his chosen people from the bondage of slavery in Egypt and called Moses to be the messenger for their freedom. God could have destroyed the Egyptians and spared the Israelites as he did with Lot in Sodom and Gomorrah, but rather showed a pattern of increasing restraint and bestowing of authority. First, God chose to communicate to Pharaoh, through Moses, instead of telling the Israelites to take shelter from an impending doom. Second, he sent nine plagues to compel Pharaoh to release his people without violence. When Pharaoh refused to obey God, he authorized the firstborn child of every household which did not honor God to be killed. By military standards, this could be called escalation of force.

Egypt was blessed to have chances to avoid death. Pharaoh's hardness toward God secured Egypt's punishment. God brought the Red Sea down on Pharaoh's army, killing them in pursuit of his people. Violence on behalf of God was done as a consequence of sin and strategically in relation to his people. There is a duality to violence. It was destruction for Egypt but deliverance for his people. God then positions the Israelites to develop his character by the removal of external influences. In order to construct a nation that reflects his character and enter into the promised land, they had to be stripped of their former nature and become warriors. The wilderness of the desert is not only a place to develop intimacy with God but also the training ground to develop his character.

Exodus 13:17 illustrates the strategic position of the Israelites: "When Pharaoh let the people go, God did not lead them by way of the land of the Philistines, although that was near. For God said, 'Lest the people change their minds when they see war and return to Egypt.'" The Philistines were warriors. They were a violent people. The Israelites were slaves and no match for any enemy. The desert was a place for them to become warriors. They had to develop the way of warriors before they could succeed in violent conquest. Their victories were gained through a process of refinement and discipline.

Violence is arguably the lowest form of warfare for the spiritual enemy. His primary methods are deceit and confusion. He doesn't need to defeat us if he can distract us. So long as we are on the path of no return momentum will do the rest (James 1:14-15). The enemy only needs to set us in motion. Divine violence was a swift method of destroying an enemy's stronghold. Correction is a call to repentance and realignment to the path of righteousness. The Israelites received punishment for disobedience because they were God's chosen people. They were also built up to enforce that punishment on nations who opposed God. As they became civilized and orderly, they were more than just light to the darkness–they became enforcers.

From Noah to Moses, Joshua to Samson, Saul to David, God's chosen people held the authority to enforce God's wrath in the form of violence. God is a warrior and judge who holds humanity accountable. Still, God is working through humanity, not around humanity. He is not a disconnected deity. When someone's violence is done for their own will, they are not a reflection of his image. Instead of removing our capacity for violence, God pours out more of his power into us in order to temper the impulse to deviate from his character.

Mankind is made in the image and likeness of God. Exodus describes God as a warrior. God's desire is for us to develop his character, including his warrior nature. The Psalmist comments, "I praise you, for I am fearfully and wonderfully made." (Psalm 139:14). God created us with an extraordinary responsibility and purpose. To be made in his image means we have a powerful capacity to harm. Yet, God tells us to be holy because he is holy, to act in accordance with our design. We all have a warrior nature within us, but warriors are violent for a purpose.

Everything Scripture tells us about God is also in the nature of Christ. If the Lord is a warrior then Jesus is a warrior who fulfills the will of the Father. The attributes of God and the character of Jesus are not in conflict. The Bible gives us a comprehensive account of God's authority and love through both violence and peace. The capacity for war and peace coexists in the person of Christ. Likewise, they also exist within us. Some argue that God is not loving because evil exists. The core principle behind this belief is the expectation that he should always save us from evil and suffering. Therefore, if God is love he is not loving by allowing us to suffer.

Death is the ultimate result of sin. God told Adam he would surely die if he ate from the tree God forbade. The fact that God did not instantly kill him is the first testament to God's mercy. The Bible says he is gracious, slow to anger, and rich in faithful love. It is easy to arrive at the conclusion that the Old Testament God and the New Testament Christ are two different people. To reconcile that Jesus Christ is the human image of this invisible violent God creates cognitive dissonance. But Jesus said he and the Father are one (John 10:30). We must consider what it means for Jesus to be part of the Triune Godhead in light of God's wrath in the Old Testament.

Jesus' journey to reconcile humanity began with a close community who developed his character. He granted his disciples authority and sent them out in an unconventional realm they have never experienced. In Luke 9:3-5 Jesus says the following:

> "And he said to them, 'Take nothing for your journey, no staff, nor bag, nor bread, nor money; and do not have two tunics. And whatever house you enter, stay there, and from there depart. And wherever they do not receive you, when you leave that town shake off the dust from your feet as a testimony against them.'"

He showed them that God would provide everything they would need. There would be no physical battles and no need for swords or violence. They themselves would experience violence later on in their journey, but they first needed to understand God's power to protect and provide for them. Now consider Jesus' words further on in Luke. Jesus warns them about the fulfillment of his purpose. Luke 22:35-37 says:

> "And he said to them, 'When I sent you out with no moneybag or knapsack or sandals, did you lack anything?' They said, 'Nothing.' He said to them, 'But now let the

one who has a moneybag take it, and likewise a knapsack. And let the one who has no sword sell his cloak and buy one. For I tell you that this Scripture must be fulfilled in me: *And he was numbered with the transgressors.* For what is written about me has its fulfillment.'"

The disciples believed the Messiah to come would conquer Rome as he had conquered their oppressors throughout history. This passage is not about using violence for the kingdom of God. It's about the disciples enduring violence for the kingdom. Jesus submitted himself to the violent nature of humanity for the sake of removing their sin. Through his death the law was fulfilled, God's character was revealed, and all humanity received the law of grace. The attributes of God and person of Christ appear to be contradictory. However, the will of God and work of Jesus are complementary.

The disciples missed the larger meaning of his words. Jesus used war language because it is a reality they understood. The call to arms was to put them in the mindset of spiritual battle. This type of language is used all throughout the Bible. God's chosen people came from a tradition of violent justice and strict obedience. Their lives were dependent on God's commands for survival. Even the most primitive minds understood this language. Jesus was also shifting their thinking toward the fulfillment of the law and a future ministry in the Gentile world. The Pharisees practiced religion to avoid wrath and leverage power. Jesus knew the disciples would face persecution and prepared them first for the spiritual battle.

God stepped into human history through the man of Jesus to save us from ourselves. He did not execute God's wrath in his time on earth because he came to be the ultimate recipient of it. Jesus did not need to be violent toward his temporal enemies because he was sent to bear the weight of their transgressions. Without his sacrifice, eternal damnation was inevitable. Not just for the Jews but also the Gentiles. Jesus is the fulfillment of God's promise to Abraham in that through one man all sin could be forgiven. Pointing to the physical non-violence of Christ–who lived within a specific context and purpose–does not necessitate pacifism as a Christian command. We are called to model the character of Christ in whatever we do, under the law of grace, and in accordance with our own consciences. Pacifism may be virtuous for some, while others are faithfully called to take up arms for the kingdom.

PACIFISM

All violence consists in some people forcing others, under threat of suffering or death, to do what they do not want to do.

– Leo Tolstoy

On September 17th, 1938 Germany established a para-military force comprised of ethnic Germans who lived on the border of Czechoslovakia in a region known as Sudetenland. This event began an undeclared war between both countries. Adolf Hitler ordered a recruiting program to organize troops in Sudetenland and had been planning a war with Czechoslovakia since the previous year. Within three days, the growing tensions between these nations caught the attention of the United Kingdom and France who pleaded with the Czechs to give the region to Germany to avoid war. Poland and Hungary were complicit in this request. By September 29th, 1938 the Munich Agreement was signed in which Sudetenland was given to Germany to appease tensions and prevent a European war.

Neville Chamberlain, Prime Minister of the U.K., returned to London the same day after personally meeting with Hitler declaring, "I have returned from Germany with peace for our time." The United Kingdom and France felt secure in the agreement while Czechoslovakia lost their strategic defense against German forces. In less than 6 months, Germany invaded the remainder of Czechoslovakia. Hitler briefed his Generals saying, "Our enemies are men below average, not men of action, not masters. They are little worms. I saw them at Munich." Germany invaded Poland just shy of a year from the signing of the Munich Agreement, officially commencing World War II. Chamberlain felt betrayed by Hitler's totalitarian agenda and inevitably ended up in a war he hoped to avoid.

In a CBS broadcast regarding peace deals in Afghanistan, Retired Marine Corps General and former Defense Secretary, James Mattis, advocated for a military presence in country until local stability could be established. Recognizing the Taliban's control of the area he stated, "You may want a war over. You may even declare a war over, but the enemy gets a vote." Two parties are involved in every conflict. One side can choose non-violence, but the opposition might not. The policy of app-easement only empowers an enemy dedicated to evil. The bartering of the concubine in Judges 19 is the most egregious biblical example of this failed behavior. Choosing peace is not the same as choosing passivity. We are in the fight if the enemy chooses to fight.

Physical warfare plagues the world. Beneath the surface of our physical wars is a spiritual war. The role of civil authorities in matters of faith are inseparable. Legal systems are intended to uphold moral laws. Jesus even echoes the overlap of faith and government service. He comments on the faith of a Roman Centurion saying, "I tell you, not even in Israel have I found such faith." (Luke 7:9). While Jesus may have disagreed with the decisions made at Caesar's level, he respected Roman authority. After all, it was his authority. He never spoke against those who were given the authority to enact justice. But he was not a pacifist. He opposes humanity's perversion of justice. He had the authority to be violent, even the power to conquer Rome, but it was not the will of the Father. The Apostle Paul clarifies the proper civil authority in Romans 13:1-2:

> "Let every person be subject to the governing authorities. For there is no authority except from God, and those that exist have been instituted by God. Therefore whoever resists the authorities resists what God has appointed, and those who resist will incur judgment."

Governments are the prescribed replacement to civil order after the fulfillment of the law. God's authority to carry out justice with violence is with those appointed and disciplined. The Torah (first five books of the Bible) give an account of the development of Israel as a nation and the establishment of his covenant. Remember that offensive violence is granted to the warriors of Israel. Many bloody battles are won for God's glory. During Christ's life, culture shifted into a secular governmental body. Paul continues in Romans 13:3-4:

> "For rulers are not a terror to good conduct, but to bad. Would you have no fear of the one who is in authority? Then do what is good, and you will receive his approval, for he is God's servant for your good. But if you do wrong, be afraid, for he does not bear the sword in vain. For he is the servant of God, an avenger who carries out God's wrath on the wrongdoer."

There is a parallel between Jesus' words in Luke 22 and Paul's words in Romans 13. Authorities who deviate from what

God has ordained become oppressors. Christ taught his disciples that they would face persecution for their faith. Jesus knew the world, including governments, would persecute Christians. At the same time, he still upheld the authority of governments to use force as there will be a final judgement to come. Leaders and authorities are held to a higher standard for a reason.

Governmental bodies are the institutions which maintain the authority for offensive violence. Pacifism on the other hand is the abdication of any use of force against another person. One understanding is that pacifism is the opposition to violence as a means of resolving disputes. A deeper explanation by the American Heritage dictionary states, "Such opposition demonstrated by refusal to participate in military action." This gives birth to the idea that Christians have no place in military or law enforcement due to the common use of deadly force.

People refuse to participate in military action for many reasons. Many people have no interest in military service but understand its necessity. However, a danger exists with those who think that no Christian should participate in a violent capacity. This belief, if fulfilled, would result in a completely secular system of government void of God's character with the authority for violence. Overarching pacifism is a denial of justice. Christian men must know how to be violent and contextualize the difference between justice and mercy. Otherwise evil will have its way in the world.

Pacifism might sound like Jesus' response, and maybe it is for some people. But a society will crumble if all its members seek peace through non-violent action. No amount of advocacy will thwart a determined enemy. The man who set out to kill me and my family tortured and killed his roommate in his apartment over the loss of drugs. How much evil awaited us at the loss of his family? We will fall to the will of an enemy without those to stand in the way with violent force. Christians are not called to tolerate evil, and evil is often fond of violence.

The effects of pacifist ideals are present today. By pulling troops out of Iraq prematurely, the country was left to defend itself with an ill-equipped police and military force that fell to a brutal extremist caliphate. In January of 2014, the city of Fallujah fell into ISIS control. In June of that same year, they further seized Mosul and Tikrit, quickly and easily gaining territory in the absence of US military strength. By mid 2015, ISIS gained control of the Anbar capitol city of Ramadi. And as if watching a rerun of the botched Iraq withdraw, Afghanistan suffered the same blow. The fight doesn't end when we choose not to fight. The spiritual enemy is not backing down, therefore the physical enemy is someone we will have to contend with.

Ephesians 6:12 states, "For we do not wrestle against flesh and blood, but against the rulers, against the authorities, against the cosmic powers over this present darkness, against the spiritual forces of evil in the heavenly places." We are engaged in spiritual warfare even if we never experience a violent encounter. Jesus did not kill anyone during his life on earth but is still a member of the Godhead. He was there at the creation of the earth, he was there at the Flood, and he is the author of the Mosaic law. The ethical problem we face is not violence; it is sin.

Violence is necessary, which is why Jesus spoke of the authority of civil servants. Violence is appalling, but sin will sear our consciences, sending us down the path of no return. It wasn't the tens of thousands of men David killed that seared his conscience, it was his sin against God. We are not all called to fight, but we are all in a fight. And a select few will be called upon to fight the wars of the flesh. For those called into the service of arms, developing the character of Christ includes the resolve to enforce God's wrath.

APPLIED VIOLENCE

Beware that, when fighting monsters, you yourself do not become a monster...For when you gaze long into the abyss. The abyss gazes also into you.

— Friedrich Nietzsche

Governments were established with a purpose. Likewise, the Church was established with a purpose. It is not the Church's role to bear the sword but the State's. Moreover, the use of deadly force to protect life is still authorized outside of government service. Exodus 22:2-3a mentions the use of deadly force in the context of self-defense: "If a thief is found breaking in and is struck so that he dies, there shall be no bloodguilt for him, but if the sun has risen on him, there shall be bloodguilt for him." In the United States, the right to life is a fundamental human liberty recognized and upheld by our Constitution. Our government permits the use of deadly force in order to preserve life, but it should be realized we've already been granted such authority by God. What is special about the United States is the authority to bear and use the "sword", meaning firearms, for such purpose. Such authority should not be taken lightly. It is one of great responsibility and reward.

Judicial violence protects us from the wolves who threaten the flock. The shepherd is not a passive guardian and doesn't work alone. Our military and law enforcement communities are those commissioned to protect the flock. They answer to the shepherd alone. What a blessing to know we have a God who understands our nature and works through us to bring about peace. Christ stood in our place as a man of action, with relentless love, who stared down the demons with an un-

223

breakable spirit. Violence on behalf of the peacemakers is an act of compassion at the risk of losing their own lives. They endure the enemy's violence for the sake of the flock. Suffering takes something from all of us, but endurance through suffering brings hope. Hope is what sustains us through the valley of the shadow of death.

Remaining unstained by violence is nearly impossible. We are naturally in conflict about the authority given to us because every human bears the image of God at our core. We were not built with a desire to harm. Through sin however, we develop the taste for it. It did not take long for the first murder to occur once sin entered the world. Non-violence is an honorable pursuit, but it is not possible unless all humanity chooses it. Until then, we must learn how to be effective warriors in order to protect the flock. Overcoming our civilized resistance to kill is not an easy task. It requires both the internal and external rewiring of a person to be effective.

In his book *On Killing*, Lieutenant Colonel Dave Grossman covers the topic of non-violence in the context of soldiers firing their weapons in war. He states that the willingness to shoot at an enemy combatant was present in only 15-20% of soldiers during the Civil War. The after-action reports found nearly 30,000 rifles on the battlefield, of which 90% were still loaded. Grossman concludes that these soldiers did not enter the battlefield with the intent to kill. Once military commanders discovered this failure to fight, training was revised to overcome a soldier's resistance to kill. Specific changes in combat training improved an individual's willingness to engage the enemy and firing rates increased to 55% during the Korean War and 90-95% in Vietnam. Even on the battlefield there are conscientious objectors.

In our homeland we see examples of the resistance toward

violence in the form of psychological distancing. A shooting occurred at the University of California Santa Barbara in 2014 that claimed the lives of 7 people and wounded 13 others. After the shooting, school official Janet Napolitano said the shooting was, "almost the kind of event that's impossible to prevent and impossible to predict." Her comment displays a common misunderstanding of human nature. Elliot Rodger, the gunman, exhibited warning signs for over five years leading up to the shooting. In at least twelve specific incidences, Rodger telegraphed his intention to harm.

Violence places us in a state of denial. The term "Bystander Effect" is a social phenomenon in which people witness a traumatic event and yet fail to act. The term comes from the 1964 murder of Kitty Genovese who was attacked outside her home in Queens, New York. Thirty-eight witnesses to the event failed to act, thinking nothing of the event. Uninhibited violence will not stop until it has consumed everything in its path. God flooded the earth when human corruption had consumed all but Noah and his family. We have an unequivocal responsibility to combat evil.

Natural aversion toward violence is not an excuse for passivity. It is easy to excuse cases of extreme violence when a lack of remorse is evident. Dehumanization of the enemy is common in warfare because it amplifies the enemy's evil and diminishes their humanity. This behavior is callous and lacks empathy. Whether genetic predispositions or social failures lead to this level of decay, Scripture tells us that we were not created this way and should never lose sight of the hope we have. Violence without remorse comes from a series of sinful choices. The enemy consumes and inhabits the willing. Losing the spiritual battle results in physical warfare.

I learned to hate my enemy in Iraq because they were killing Americans. Though I never questioned why they were killing Americans. I just believed they hated us because we were their enemy. Violence is easy when it's driven by emotion. But not everyone trying to kill us was a legitimate enemy. We don't have to hate our enemies to rightfully use deadly force. Hatred is a self-justifying emotion that skews the lines between murder and killing. I've seen many people fail to rationalize their actions in order to relieve themselves of the guilt of violence. Even justified violence weighs heavy on the heart. Hatred is the easy way out because it is driven by emotion. When the smoke clears, hatred betrays us and we are left to reconcile our rage.

In times of my own suffering I fall hard into prayer. At times, I've confessed to God there have been moments I just want to hurt people. Anger consumed me. I understood the context for violence but the distinction between justice and vengeance was a blur. There were moments that felt like God responded saying, "I know how that feels, and I've pulled the trigger too. Generation after generation I've sent men to their graves. I've sent women and children along with them. But it did not change their hearts. Their evil was beyond correction and destroying the whole world won't remove the darkness of the human condition. So, I sent my Son to love them in their anger and to forgive them in their pride. When the world was doomed, my love was the only hope they had for life." When I'm losing a spiritual battle, prayer is all that will lift me out.

There is something uniquely powerful in Christ's gift of the cross. Christ expresses the depth of his suffering to his disciples saying, "My soul is very sorrowful, even to death." (Mark 14:34). When Christ gave up his life for humanity, it was the utmost expression of love because it was paid to an undeserving audience. Jesus is the resolution that enables us to submit to his authority and the governing authorities to avoid the path of no return. Violence can stop violence, but it cannot end the evil in man's heart—it can only judiciously keep it at bay.

Jesus is deeply familiar with the violence of our world. He poured out his love with unwavering empathy beyond our chaos. He is telling us it is not a sin to be angry, but to not let bitterness consume us. Mercy and grace are meaningless without justice and judgement. We are loved despite our anger, our violence, our deceit, and Christ has proved that to us. Romans 5:8 states, "But God proves His own love for us in that while we were still sinners, Christ died for us!"

We have forgiveness for our sins but governments still determine civil punishment. Those who are called to be civil servants must be faithful with the authority to use violence. The only thing stable in an absolutely mad world is God's love. It is his love that put everything in motion and pierces the hearts of mankind. God enables the peacemakers to be enforcers of the light. In everything we do—violence or peace—we are still conduits for his love. God's character is observed through humanity, both his judgement and merciful love. If good men remain idle the enemy will cast their vote. Likewise, victims will lose hope. God will restore victims of injustice even if that includes violence. The kingdom of heaven is suffering violence and the war will not be won with tolerance and pacifism.

CHAPTER FIFTEEN
CONFRONTING THE MONSTER

"Fancy thinking the Beast was something you could hunt and kill!" said the head. For a moment or two the forest and all the other dimly appreciated places echoed with the parody of laughter. "You knew, didn't you? I'm part of you?"
– William Golding

A FIRE SWELLED INSIDE me during my first semester in college after the Marine Corps. The crushing weight of insignificance crept in. I felt void of purpose, surrounded by college students who were disconnected from the realities of life. I began to blame them for the feeling of isolation as they were unable to relate to my suffering. I pressed on through course after course, trying to accept the season of life as students complained about early class times and inconvenient responsibilities. I grew furious and self-righteous in my anger. Their lack of resilience for anything challenging was pathetic. My expectations for community college were already low, yet this felt like a personal attack.

I frequently suppressed the urge to scream at people. Every student who walked into class late or asked for an extension on an assignment was at risk. They seemed so lazy and undeserving of grace from the instructors. Pride fueled my frustration. I felt the need to display my own accomplishments to be seen—and to belong. The other part that angered me was witnessing a generation filled with potential wasting their time. They carelessly squawked at politics, advocated for safe spaces, and loitered their way through education. A student ID was the only thing we had in common. I felt I had achieved so much, and yet there I blended into a crowd of entitled children without direction.

I sat unnoticed in the back of classrooms while the loudest voices were awarded grades for participation. My capacity for aggression was tested almost daily. I fought against my own bitterness, knowing other people had no responsibility to facilitate my recovery. I was in desperate need of a new mental framework. The social change in my environment fractured my pillars. College was my first integration into a community with a lack of shared experiences. The communal pursuit of a common goal did not exist. No matter how friendly I was to people, it failed to achieve what I had in the Marine Corps. An existential crisis of meaninglessness buried my soul.

I was intimately familiar with the dark side of life, using bad language and harsh humor as coping mechanisms. The thought of submitting to the fragile nature of others was contemptible. The "safe" environment of college was meant to keep people like me away from other students. Relating to other students meant having to sugarcoat the details of my experiences so as not to "sear their consciences." The threat of isolation loomed unless I conformed to the soft ideals of a protected society. Many were against the war in Iraq, believing it was morally wrong. People thought Operation Iraqi Freedom was a crime against humanity and participation meant complicity. Death and suffering dominated the media, and we had our hands in it.

The media proudly displayed the Marine Corps' violence. It was our job. It was just business. When conversations about violence or tragedy arose in class, some students would publicly denounce it as evil with haughty political references. Few understood the intricacy of violence. Many advocated a pacifist stance. However, the disapproval of students revealed something deeper. It was the inability to distinguish justified violence from criminal violence. Being violent for the benefit of others was incomprehensible to them. I had seen in myself a truth they denied: Every human being has the capacity for violence no matter how much we suppress it.

HUMAN NATURE

Circumstances don't make the man, they only reveal him to himself
— Epictetus

I was having a debate about firearms with a professor during class one day. "I could never harm someone." He jabbed. "If someone broke into my house and tried to hurt me or my family, I wouldn't use force much less shoot them." He proclaimed to be a pacifist, which challenged my understanding of human nature.

"So, you're saying you would not harm someone to prevent them from hurting your family?" I asked.

"Yes." He replied. "I'll never own a gun. I don't even have the ability to hurt someone. It's just not in me."

"Oh yeah it is," I said, "You don't need a firearm, but someone threatening your family is the most primitive context that can bring out your ability to be violent. And if it doesn't, your wife's maternal instinct will deal with the threat."

I met more anti-violence advocates on college campuses than I have anywhere else. Pacifism produces nominal realities that do not prepare us to confront reality. The ultimate theory is that by choosing peace the world will become more peaceful.

We are inherently sinful, which means God's grace is the only thing that brings true peace. However, passivity is not a form of justice simply because it is 100% mercy. I can't image the insecurity this professor's children might feel knowing their father willingly exposes them to the dangers of the world because of his moral high ground. People need more than to just *feel* safe. The illusion of safety is destroyed when the enemy shows up.

Psychological separation from our capacity for violence is a privilege of civilized society. The civilized are shielded from the realities of uninhibited human nature. We are doing ourselves a disservice by refusing to be educated on violence. Some believe violence is an intrinsic trait of the outliers of society. Granted, people don't wake up and decide to start killing people. There are some who are genetically predisposed toward anti-social behavior (as it is academically classified). Social factors play a role in the manifestation of what we would consider "evil" behavior. To reduce violence to a rare genetic quality is to deny a deeper knowledge of ourselves. Violence is a choice, not a product of a different nature.

Hope is found in believing that the Spirit of God can transform a heart to desire harmony for the world. Still, believing the world will only be transformed this way is vain. Our role includes defending the flock. I hesitate to listen to advocates of non-violence when their argument comes from emotion. Desiring peace is virtuous, even admirable. Granted, physical harm is more than an attack on another human. It is an attack on God because we bear his image. The spiritual enemy is at war with God. He encourages us to kill one another as a means of attacking God. Remember that God flooded the earth to protect a righteous man and his family. He was protecting his image from the enemy.

No matter how much we morally oppose it, we all have the capacity for violence. The vice of a protected society is that most people are sheltered from the experience of pure corruption. Every society is susceptible to spiritual corruption even if it does not manifest physically. We can be non-violent and still lose our souls. The grand illusion of civilization is that we can become spiritually bankrupt while existing in physical harmony. We don't need physical violence to know that the world is headed for destruction. Avoiding conflict does not constitute an inability for destruction and becomes meaningless if our souls are lost.

We should care more about our proclivity to reject God rather than our physical manifestation of violence. It is said that people end up in hell for their disbelief not simply their sin. Spiritual decay always precedes physical decay. We were created in the image and likeness of God, but our fallen nature leads us away from our eternal design, declaring war against it. We are either moving toward righteousness or sin. The descent toward human destruction began with separation from God. The further we drifted from God, the more violent we became. Disobedience to God's commands may seem harmless until it matures into something unspeakable. The pathway to violence originates from the heart. The spiritual enemy must be dealt with first and foremost. Civilized or not, we live amongst enemy territory on this planet where the devil prowls around like a lion.

Enemy territory is anywhere God is not being honored. Self-serving violence indicates poor spiritual health. The kingdom of heaven is under assault. Threats are an inescapable component of life. Our ability to construct civilizations and thrive on the collective efforts of each other is proof that we can manage the violent nature inside us. But we cannot contain violence permanently. Violence is a heart issue. The inward will manifest outward. Murder begins in the heart, therefore the enemy's ability to take ground begins in the hearts of men. If we are not honoring God with our heart, soul, mind, and spirit then violence is an inevitable reality awaiting us.

Humanity shares similarities with the animal kingdom in that we both have a primitive drive for survival. One thing that separates us is our capacity for malevolence. Animals do not harm other animals to satiate emotion. Physical threats pose a risk to the preservation of society. Yet with humans, things like selfishness, greed, and lust turn us against each other. Without God these desires grow beyond our basic needs. The chasm of the soul craves more and more. Violence breeds more violence as sin gives birth to sin. Humans are social creatures and hierarchal by nature. The moral corruption of an individual can become the moral bankruptcy of a nation. Mass genocide historically originates from the moral decay of a strategically positioned individual. The tyranny of a nation can be traced to the enemy's stronghold on the human heart.

Leaders are held to a higher standard because they hold authority over others. The enemy attacks prominent figures as a conduit for collateral damage. Authority figures become the catalyst to blind the masses from God. A catalyst is one who awakens the dormant and activates their physical capacity for harm. Most people who are aversive toward violence are simply dormant. Hitler was a catalyst. Stalin was a catalyst. Mao was a catalyst. These catalysts manipulated the dormant to fulfill their violent grievances. These leaders were subtle with their malice until they were insulated from repercussions. Then they became overt in their destruction.

Prominent violent catalysts are easily identified by the amount of damage they cause. Smaller scale catalysts fly under the radar until a heinous act has already been committed. In the

1999 Columbine High School shooting, both shooters exhibited warning signs that went unchallenged prior to their attack. Yet, Eric Harris was a catalyst for Dylan Klebold. Klebold was considered highly intelligent and attended his senior prom shortly before the shooting, though he struggled with insecurities and feelings of worthlessness. Klebold sought the approval of Harris, which postured him to be manipulated by the psychopathic catalyst. Without Harris, Klebold likely wouldn't have manifested his capacity for violence at Columbine High School.

These catalysts were not loners, as some might assume. They held jobs, went on dates, and had many friends. Harris was considered an intelligent psychopath with a God complex. Both Harris and Klebold had a fascination with weapons. They had a history of felony theft and juvenile correction programs. They consumed violent movies and video games. Harris showed a fascination with other school shooters, wrote a paper about a school massacre for his English class, posted plans on a website, recruited a fellow student to acquire a firearm, and wrote explicit plans in his yearbook. The warning signs exhibited by Harris and Klebold unraveled over a year leading up to the shooting. They were observed by students, faculty, and law enforcement who took no action against them.

The pathway which leads to violence is complimented by passive action. The media they consumed influenced their descent toward destruction. Both were fascinated by the video game Doom. Investigators observed "reloading" behaviors that mimicked character motions from the game when viewing security camera footage of Harris and Klebold. They were also obsessed with the movie Natural Born Killers, using "NBK" as a code word for the massacre. A majority of active shooters tend to exhibit what is called "leakage" before a violent attack.

Additionally, Harris and Klebold were not the only active

shooters who digressed the way they did. At least 74 copycat active shooters cited Columbine as a motivator for their massacres, planning killing sprees on the anniversary, using the same code names and weapons, and consuming the same media. Twenty-one of them acted out their plans. Consumption of violent media becomes a perpetuating cycle. The Columbine shooting inspired numerous movies and television shows. Violent media influenced Harris and Klebold, who influenced media, which influenced others. A video game replaying the acts of Columbine was even created by a professor of Film and Media Arts. The professor stated, "It was a bit scary, once I learned more about these boys, because it was like I was looking in the mirror and I didn't want the same fate for myself." Self-serving violence is a catalyst for more violence.

Seeds of destruction are sown long before they manifest in the world. The spiritual enemy is not always overt. Violence begins by listening to the enemy's voice over God's. If the dormant can be activated by a catalyst, then we are all vulnerable under the right conditions. We are all capable of descending into chaos. Satan was the catalyst who corrupted one-third of heaven's angels. He is the conduit for which destruction is mobilized. His desire to compete with God expelled him from the heavenly realm. One catalyst activated a third of the heavenly hosts from a dormant posture. If sin is not contained, then humanity will terminate its own existence. Stories of human corruption saturate the Bible but they also testify to God's righteousness.

Christ came to restore our vast separation from God. Jesus' death on the cross removed the grip of sin on the human heart. His earthly purpose was not to end wars but to destroy the enemy's stronghold on the heart where all violence begins. The

Romans were known for their brutality and disregard for human life. Violence was even a form of entertainment. People were intimately familiar with the power of evil. Jesus often spoke of the battle between heaven and hell using war language. Combatting humanity's destructive nature is not only done by the sword but with the sword of the Spirit.

God's law has been written on our hearts and radically informs our consciences. The primal nature of humanity matured in conscience though we have not lost our propensity for violence. Warfare is strategic, and Scripture is our battle plan for engagement. Humans don't need to be taught how to be violent; we only need to be disconnected from God and the enemy will move. True civility is the ability to be violent yet choosing peace until violence is necessary. The threshold of where that necessity lies is ill-understood by the passive. We must come to terms with the violent potential we all possess before a catalyst comes along to exploit it.

FREEDOM OF CHOICE

For even among them, some refused to kill and others stopped killing. Human responsibility is ultimately an individual matter.

– Christopher R. Browning

Freedom is an important topic of faith. It is often refuted by other religions as well as the non-religious. Because so many have difficulty reconciling a violent world with an all-loving God, a discussion about our own power to alleviate suffering in the world is common. But as we fail to achieve the ideal resolution for suffering, one conclusion is that we really don't have the power to solve it, much less choose a side. It's easier to deny our ability to fix the problem or share any moral responsibility in causing it. This is essentially the position of

236

determinism. I can understand the motivation to believe this, but I can't believe that we have no power to affect our lives and those around us.

One observation is irrefutable—we all have a desire to alleviate suffering in the world, even if it is just our own. However, we deviate on the best methods for alleviating suffering. Likewise, we probably differ in our beliefs about the root cause of suffering. Believing in free will assumes taking responsibility for our actions in the world. I'm not saying that our attitudes on a bad day are going to destroy the world, but our actions add to the grievances of someone who might be on the pathway to violence. Remember that a student's rejection to attend art school contributed in a small way to the snowball that became a second world war. Concluding that a harsh comment alone will lead to the death of another human is reckless. It's possible but highly improbable. How we treat people affects their emotions and decisions and we are held accountable for it.

When we choose to be a light in the world we may never know how it affects someone. A hostile attitude tears down the fabric of someone's reality while a hopeful attitude can transform life for the better. Recently a friend called to tell me that he and his father talked someone off a bridge near my hometown. Our words have the power to give and take life. On the other hand, we have a fallen nature and are capable of great harm. The daunting reality of violence in the world makes it hard to accept that humanity chose this way of living. Most people have not experienced the depth of suffering it takes to act out violently.

But what about the good we see? Without the freedom to choose, all justice lies outside of us. If we are a product of a deter-mined world without free will, then our actions are a logical string of events and not something to be faulted. We shouldn't feel any emotional weight for the way of the world if it is just following a natural progression of events. The reality is that we do. The law has been written on our hearts and our consciences impact our choices. Denial of free will often results in a Darwinian view of the world where natural selection dictates behavior.

Like animals, we are built with the drive to fulfill our basic needs at an unconscious level. Have you ever noticed how you throw your hands up to protect your head when someone throws an object at you? The Limbic System in your brain is programmed to deal with threats. It instinctively causes us to throw our arms up to protect our control center. Our physical preservation is so ingrained in our hardwiring that we act on it subconsciously. That makes a strong case against free will in lieu of our primitive nature since we are naturally solipsistic. But morality is always a conscious choice.

When you look at animals, it's easy to see the survival of the species. It's not a coincidence that the animal kingdom is filled with extreme violence. Yet, the struggle we have with this suffering rarely ever leads to suppressing the nature of wild animals. Animals will do anything necessary to preserve their lives. Violence is a product of utility. They are not capable of abstract thought. Morality is uniquely human. Therefore, the problem we have with violence is a heart issue. We aren't fighting to eradicate violence in animals like we do with humans. The only species on earth capable of malevolence is humans. This kind of violence doesn't manifest in the animal kingdom. Humans can harm for sport, while animals harm out of necessity.

Supererogatory acts are also an interesting challenge to determinism. These acts are exclusive to humankind. Super-erogatory acts, such as someone sacrificing their own life for another, go against our survival instincts. Determinism is rooted in biological programing as an explanation for all human behavior, yet fails to adequately explain this phenomenon. Only the Spirit of Christ in us would lead us to give up our own lives for someone else. Sacrificially enduring pain, even to the point of death, and not doing anything to change it is the ultimate rebellion against the flesh. To suffer for Christ is to willfully choose life.

The earliest example of free will is seen in Genesis 2:16-17. Here, the first man on earth (Adam) is given the option of choice through a directive. It says, "And the LORD God commanded the man, saying, 'You may surely eat of every tree of the garden, but of the tree of the knowledge of good and evil you shall not eat, for in the day that you eat of it you shall surely die.'" Adam can follow the order or disobey it. Sin has not entered the world, so pain and suffering do not exist. Nor does Adam have any knowledge or experience of death. Intimacy with God was uninhibited by sin. Any motivation to act from a survival instinct is null when death has not yet entered the world.

The book of Deuteronomy also demonstrates humanity's ability to make choices. Deuteronomy is considered the "second law" of Moses which recounts the Israelites' forty years in the wilderness. As they prepare to enter into the promised land, they are given three different speeches through Moses from God regarding the laws which have been established for them. The specific words used in this text reveal the nature of the relationship between God and creation, namely that of their freedom to choose. Most commands are a direct call to action. One example comes from Deuteronomy 11:1:

> "You shall therefore love the LORD your God and keep his charge, his statutes, his rules, and his commandments always."

239

The Israelites are not told what they *will* do but what they *should* do. The word "If" appears nearly seventy times throughout the book of Deuteronomy as a conditional word. In chapter 11:26-28 it says:

"See, I am setting before you today a blessing and a curse: the blessing, if you obey the commandments of the LORD your God, which I command you today, and the curse, if you do not obey the commandments of the LORD your God, but turn aside from the way that I am commanding you today, to go after other gods that you have not known."

Free will is the most dangerous gift a Creator can bestow upon the created. It bears the weight of responsibility. It carries the duty of violence. It requires a considerable amount of trust. Freely choosing to follow God is the act of stepping out into a great unknown. God knows what will happen, but it is clear from the text that there is a choice to be made. God told the Israelites what he would do in response to their choices. One verse that shows God's knowledge of Israel's future is Deuteronomy 31:16. It says:

"And the LORD said to Moses, 'Behold, you are about to lie down with your fathers. Then this people will rise and whore after the foreign gods among them in the land that they are entering, and they will forsake me and break my covenant that I have made with them.'"

God knows *what* will happen and *why* it will happen. Deuteronomy 31:21b states:

"For I know what they are inclined to do even today, before I have brought them into the land that I swore to give."

Free will means we have power to affect change in the

world without knowing exactly how that change will play out. God promises that if we choose him it will work out for our good. It is the blank check to God that says, "I don't know where this is going, but I will follow because I choose to believe you are who you say you are." Relationships are built on trust. Spiritually, this is called faith. God's ultimate desire is for humanity to be reconciled to himself. Relationships are grounded in the mutual exchange of love. Love is something which cannot be taken from us against our will. In Deuteronomy 13:3b God makes his purpose clear:

> "For the LORD your God is testing you, to know whether you love the LORD your God with all your heart and with all your soul."

So, what is love without the freedom to choose? What is choice without the risk of trusting? I didn't join the Marine Corps because of an assured safety or benefit. I joined because I couldn't live knowing there were problems in the world that I could do something about. I refused to be idle. I knew the progression of natural events would escalate to worse problems if they went unchecked. I first chose to believe in God. My desire to serve God inspired me to do needed and timely work in the world at the potential expense of my own life, but with the promise of his peace. We are not aimless beings behaving as pre-programmed and unaccountable lifeforms. We are image bearers in need of reconnection with the source of life. The spiritual war we are in is a matter of freedom. Sin is crouching at our door, and we must *choose* life.

NEED FOR AUTHORITY

But the people refused to obey the voice of Samuel. And they said, "No! But there shall be a king over us, that we also

may be like all the nations, and that our king may judge us
and go out before us and fight our battles."

– 1 Samuel 8:19-20

What is it about the human condition that desires a leader? Humans are hierarchical by nature. We always look for someone to show us the intended version of ourselves. The human need for authority reveals our eternal design. History is filled with nations ruled by kings and appointed leaders. We love sending someone else to do our bidding. If they fail, then it is not our failure but someone else's. We believe we can always start over and elect someone new. But the problems they cause, we inherit. No matter how much we abdicate our responsibility to someone else, we still suffer the consequences of their actions. We need an authority figure, but anyone who reigns without God's anointing will increase the troubles in the world.

God brought the Israelites out of slavery and declared that he would be their leader. Even before their request for a king was made known, he understood their need for authority. The God who created them in his image and likeness, the source of complete fulfillment, heard their cry of oppression. In response, he brought Moses to be the conduit for his authority to be received. He allowed them to suffer in Egypt to be humbled into recognizing their need for a savior. Still, the overwhelming mercy he showed them did not compel their hearts to follow him. They lost their desire for God when their oppression ceased. Although they were free, they complained, stating they were better off as slaves in Egypt.

Humans will look everywhere except to God for solutions to their problems, including returning to old methods that don't work. Israel wanted a king who fit their image of an authority figure. They wanted a king to appease their sensibilities and satisfy their desires. God tried to show them another way. They weren't impressed with the God who led them out of Egypt because he offended their sensibilities. They were hard hearted. They wanted a physically present ruler that would fight their battles. Instead of trusting the Creator, they placed their faith in the created.

Ancient Israel's demand for a king is no different from how we behave today. "If only our leaders would lead by example," it's been said. The Israelites abdicated their strength to a human authority, a king who fought on the frontlines with them. In our modern world, those declaring war are not the ones who participate in it. Our leadership is dramatically different from the kings and kingdoms of ancient Israel. Kings led their nations into battle. The Israelites were wrong to demand a king, yet they had the culture of leading from the front. We wrestle with violence because we elect authority figures who advocate pacifism rather than faithfully guide us through the valley of the shadow of death. It takes a warrior aligned spiritually and with integrity to properly guide the flock.

Humans hedonistically idolize what make us happy and condemn what brings pain. We excuse ourselves from God-given responsibility because it releases us from ownership of our problems. We blame individuals for bad leadership and in-stitutions for corrupt business practice. We condemn elected officials for their misrepresentation of our values. Then we go on electing others to take their place in a never-ending cycle of failure.

While God appoints leaders in government, they are never replacements for himself. God is the ultimate source of our healing and knows we cannot achieve perfection. He tells us to pursue him with all our heart, and he will work through us. Confronting the evil we see in the world is not going to come from an improved policy. We go to war against unnecessary suffering when we take responsibility and stop abdicating our power to someone else. God is not calling us to elect the right leader. He is calling us to be the leaders. He entrusted us with his power. We are vessels for his grace in the world. Faith empowers us to have hope in a fallen world. In order to avoid suffering from unnecessary violence, we must understand the utility of justice and become the leaders who carry it out judiciously.

When we pray for success, we are given wisdom and resources. When we pray for healing, our spirits are tested to build perseverance. When we pray for justice, we are led to the training ground to become warriors. When we pray for peace, the opportunity to fight for it arises. God has shown us throughout history that his work will be accomplished. When Christ died on the cross the law was written on our hearts and we received his Spirit. We have been given tremendous power in this life. The victory has already been won but we must still fight the battles.

TAKING OWNERSHIP

And Samuel said, "As your sword has made women childless, so shall your mother be childless among women." And Samuel hacked Agag to pieces before the LORD in Gilgal.

– 1 Samuel 15:33

Samuel was a respected prophet and man of God. His role with God's people was to guide them in the way of the Lord. He appointed Saul, guided the nation of Israel in wisdom, and let them run their course. Their rejection of God's commands led to their disillusionment with God. When the Lord rejected Saul as king Samuel stepped in to restore order over God's people. The last person expected to take up the sword would have been a prophet. Yet, Samuel assumed temporary authority over Israel through a traumatic display of violence–mutilating king Agag publicly.

The focal point in this story is not violence, it is absolute obedience to God's commands. Saul lost his anointing because of his disobedience in not killing king Agag. When called, obedience to the Lord is more important than the preservation of life. This is often the stumbling point when the topic of violence arises in conversation. Instead of seeking a deeper understanding of how to be a faithful servant in arms, many outright deny their capacity to harm. In cases where we are entrusted to protect our allotted flock, choosing to lose our lives at the hand of evil is not martyrdom it is disobedience.

Conditions that exceed our civilized capacity can quickly descend us down the pathway of violence. Deuteronomy 28 exposes the brutal future of the Israelites should they disobey the Lord and find themselves under siege from the enemy. In this context, their civilized communities will be cut off as well as their resources. They will be oppressed and in a state of survival. Verses 54-57 state:

"The man who is the most tender and refined among you will begrudge food to his brother, to the wife he embraces, and to the last of the children whom he has left, so that he will not give to any of them any of the flesh of his children whom he is eating, because he has nothing else left, in the siege and in the distress with which your enemy shall distress you in all your towns.

The most tender and refined woman among you, who would not venture to set the sole of her foot on the ground because she is so delicate and tender, will begrudge to the husband she embraces, to her son and to her daughter, her afterbirth that comes out from between her feet and her children whom she bears, because lacking everything she will eat them secretly, in the siege and in the distress with which your enemy shall distress you in your towns."

Cannibalism was common during times of siege in ancient cultures. The verses here state, "the most tender and refined." The history of the Franklin Expedition is an example of this human catastrophe. In the mid 1800's, England was still a dominant world power. It was believed to be a model for civilized society. Their naval exploration was proud and ambitious—a privilege other countries could not afford. Two ships, the HMS Erebus and HMS Terror set sail from England in 1845 to find the Northwest Passage but were caught in an ice pack, stranding the crew for over a year and a half on the frigid ice. The ship's crew had already taken casualties over their voyage. They set out on foot, hoping to secure food and avoid death from their lead poisoned rations.

None of Franklin's crew made it out alive. Four decades after Expedition went missing, rescue crews slowly discovered traces of Franklin's crew. They found that the crew managed to survive for a time after abandoning ship. England sought to recover what it could of the expedition and give an honorable eulogy for the men brave enough to face such harsh conditions. While most died from starvation, disease, and hypothermia, the most disturbing discovery was that that crew had turned to cannibalism to survive. England's pride was destroyed at the thought that their own countrymen could digress to behaviors only seen in savage peoples.

Modern psychological research reveals that hunger is a powerful motivator of behavior. It can override fear, anxiety, thirst, and social order. We laugh in our social groups when someone becomes "hangry" from lack of food. There is something to be revered when meeting our needs. It's unwise to assume how little someone may go to get their own needs met. Violence is commonly used to acquire resources on a systemic level in countries where law enforcement is weak. We hardly notice this in first world countries because we have reliable security that facilitates our basic needs.

In the United States we far exceed our basic needs. Our level of civilization has removed us from the terror of our violent nature to the extent that when we do experience the dark side of humanity, our minds rationalize brutality as inhuman. We intellectually distance ourselves from violent individuals by calling them evil or biologically different. When we distance ourselves from others, we deny our capacity for the same behavior. Most people never commit heinous crimes, and even refuse to accept their ability to perform them. An interview conducted by Rolling Stone in June 1970 captured similar words from the deranged Charles Manson, who stated that anything we see in him is also within us; that it was dependent upon how much love we had for others. He demonstrated the uncomfortable truth that he is just a mirror to what we can become if we allow darkness to consume us.

Those words, though manipulative for their context, have an underlying reality. A tremendous amount of accountability lies in the wake of violence. The capacity for violence is within all of us. Law enforcers and military service members are no less violent than a criminal, cartels, the Taliban, or Al-Qaeda. We must not conflate those who have honorably dedicated their lives at any point to a proficiency with violence in a context where God has not bestowed such authority.

The detriment of a civilized society is the belief that violence is never justified. Such societies deny the application of justice. Those who would deny their capacity for harm weaken the flock. Evidence for this can be seen in states where governments legislate use of force from the perspective of the recipient and not those enforcing justice. Our minds are wired to protect us from threats using fear. Fear is a powerful informant, but can pervert justice when focused on the wrong variable. A society in which the majority have never fought for their basic needs is a society that will abdicate its fighting to a select few, then condemn them for using force from a lofty distance.

Evil exists because we chose it both actively and passively. Adam and Eve were given thousands of right options and one wrong choice. They chose the one option that brought sin into the world and did so from a sinless position. We lost our humanity when we lost our connection to God. We now have a knowledge of things that humans didn't previously have. Maybe God allows suffering in the world because we struggled to choose him in a world absent of it. Pain makes God relevant to us. We live with the burden of the flesh which is driven by temporary rewards and punishments. Still, God loved us enough to send his Son to die in our place so we could have freedom from the bondage of sin.

The greatest evils in the world are corrected by examining the heart. No legislation, no alliance of nations, no solution other than the Word of God can treat the brokenness of our human condition. God pursues us for reconciliation. God uses the willing to alleviate the brokenness in the world. He does not need us, but he chooses to be with us. He is our only hope for peace. If we could solve the problems we created, then we would crown ourselves with glory. Where is God if we solve poverty? How big is God in peacetime? We rarely think of him without the need for him. Our spiritual adversary need only keep us distracted to massacre the image of God. Pacifism is a grand distraction. The consequence is a comfortable society oblivious to the prowling lion waiting to devour the complacent.

CHAPTER SIXTEEN
THE ADVERSARY

From the days of John the Baptist until now the kingdom of heaven has suffered violence, and the violent take it by force.
— Matthew 11:12

JERRY'S EYES WERE FIXED on me with an unbroken gaze supported by a stable finger pointed in my direction. "If you ever harm her, I'll kill you." He said. I smirked, fighting the urge to smile while slightly confused. At his age, he was not much of a physical threat. But years of combat deployments meant he at least believed it. I listened to the rest of his concerns regarding my relationship with his daughter.

Jerry was drafted in the Army and spent his first tour in Vietnam. He then joined the Navy as a Corpsman and continued his service, completing another tour before the war ended. But the country he returned to despised his service. Like most Americans returning from Vietnam, he buried his experiences and carried on with his life. He served the rest of his career in the Navy until the late 2000's when he retired and took a job in a civilian hospital near where I lived.

A few years after I discharged from the Marine Corps, I

began dating a gal from my hometown. She grew up without a father. Jerry was the first person she looked up to. He had dated her mother for quite some time and they planned to marry but kept putting it off. Jerry and I got along well throughout the duration of our relationship. He admired my commitment to the infantry during a time of war. When my relationship with his daughter began to decline, I sat down with Jerry and expressed my concerns for our future together.

"Michaela and I are trying to work things out. I know she values you in her life. I'm here to talk outside of her presence so there are no barriers in what you have to say about us." Michaela struggled to support me in my transition into civilian life. I shared my experiences with her so she could understand my pain, but that did not resonate well with Jerry. Before I could explain my concerns about our relationship, he unleashed his own.

"If you ever harm her, I'll kill you." He set the tone immediately. "You should have never shared your journals with her! Do you realize that she is afraid of you? When I returned from Vietnam we didn't talk about it to anyone, especially not with women. If you need to talk about it, you can do it in the presence of men who were there, who understand."

He paused to calm his nerves. I could tell this issue was a personal offense. I lost my focus after hearing his position. I wholeheartedly disagreed, but nothing I could say would have changed his mind. He continued on, "If you're dealing with post-traumatic stress then get help. Talk to someone at the VA. I've been where you are. I understand what you're going through. I have a hard time sleeping at night, that's why I live on my own and not with Michaela's mother. Well…" He stopped himself, catching his unraveling emotions. "That's one of the reasons why."

I could see his own post-traumatic stress as he caved ironically in my presence. "Look, I lost my wife when I was in Vietnam. That is the hardest thing I've ever dealt with. I'll never put myself in a position to face that again. That's really...why I won't marry her mother." I was sympathetic to his pain though conflicted on his reservations. He finished his rant, holding firm to his position of isolation.

I wanted to offer my condolences, or some sort of hope, but I was sidetracked by his comment that Michaela was afraid of me. She never said anything to me directly and showed no sign of discomfort with what I had shared. Apparently, she had voiced her concern to him. His initial threat made sense now. The pain he felt all those years caused him to bury his past for the sake of the world's innocence. And here I was stealing it from his daughter. He knew what inner demons could manifest, and his way of controlling them was through isolation. He lived alone, preventing himself from having another loving relationship even if it couldn't replace what he'd lost. He valued the innocence of others at the expense of his own well-being. Instead of inviting others into his healing process, both he and the ones he loves ended up paying the price.

CULTURAL VALUES

Not till we are lost, in other words not till we have lost the world, do we begin to find ourselves, and realize where we are and the infinite extent of our relations.

– Henry David Thoreau

On September 11th, 2001 the U.S. became united on a national level. Political divisions fell silent for the first time in decades. For a brief moment, we saw ourselves as Americans aligned against a common enemy. The threat of terrorism was confirmed, and thousands rushed to join the ranks. The decision to invade Afghanistan was initially supported. But political division resurfaced in 2003 when we invaded Iraq. Operation Enduring Freedom (OEF) and Operation Iraqi Freedom (OIF) were two separate wars in the context of political ideology. For most veterans, including myself, they were the same. I signed my contract to join the Marine Corps infantry five months after we invaded Iraq. I signed knowing I would end up in one of two countries.

The war in Iraq parallels the Vietnam War. I returned home to protests, though I would never compare what I experienced to the protests of the 60's and 70's. The damage of public backlash during the Vietnam War was far greater than anything OIF or OEF veterans faced. Unlike the Vietnam War, everyone I served with joined on their own free will. We took ownership of the uniform we wore. Many Vietnam veterans did the same, but the draft pulled unwilling Americans into a conflict regardless of their convictions, then returned them home to a country that rejected them. The public voice made sure that all who wore the uniform were shamed for their involvement.

The politics involved in the decision to invade Iraq skewed the moral basis behind the war and divided public support for retaliation. This division highlighted the polarity of cultural values. On one side were those who saw physical force as the appropriate response to a terrorist attack. On the other side were those who advocated every possible non-violent approach. Those who supported a military response to the 9/11 attacks understood the logical progression of terrorism unchecked. Opposition to violence was seen as weakness, where non-violent advocates saw their position as an expression of strength.

Still, many supporters and anti-war advocates saw the nefarious political agendas underlining a full-scale military response. The protests of OIF differed from the protests of OEF. When it was discovered who was responsible for the terrorist attacks on 9/11, the majority of the country favored a response in Afghanistan (OEF). Many Americans felt their anger was manipulated when the decision to invade Iraq (OIF) was made.

Roughly a year after I returned home from the military, I was having coffee near a school I planned to attend. I started a conversation with a gal about her experience at the local university, which led to a few dates. She began inquiring about my experiences overseas. The conversation shifted when her underlying beliefs surfaced:

"How can you be in the military and call yourself a Christian?" I was caught off guard by her statement. "I would never join the military. I don't see how you can be in the military and be a Christian," she followed.

"Are you saying that it's difficult to be a person of faith and serve in the armed forces or that you find it a contradiction for a service member to claim faith?" I responded.

"It's a contradiction…" she ended.

Her moral objection to the military was rooted in the belief that killing under any circumstance was wrong. She assumed that being a Marine automatically made me a killer. I was immediately labeled, convicted of the crimes of a group, and shamed for believing in God at the same time. The most offensive thing about her sentiment was not the personal accusation, but the limited view of God's character. I rejected her shallow conviction that I was unqualified as a Christian simply because violence appalled her.

There are still those in the Christian community who condemn military action, or at least Christian participation in it.

The Cambridge English Dictionary defines pacifism in the moral sense as, "The belief that war is wrong, and, therefore, that to fight in a war is wrong." In a religious context, *wrong* implies that something is sinful. Pacifism has been a doctrinal mandate in various denominations of Christianity for centuries. The debate about Christian participation in warfare is not a new one.

Christians who support pacifism often make references to the character of Jesus in the New Testament as a means to advocate non-violence. Prominent pacifist theologian Greg Boyd for example, asserts that people are wrong to believe God is responsible for any violence. In his view, Christians are wrong to be involved in warfare completely. He insists on defending God from a violent image despite God claiming responsibility in many instances. According to Boyd, the Bible states that God will judge sin but that judgement was solely carried out at the cross. Pacifism can be maintained by some Christians but Boyd is attempting to reinterpret Scripture through his own sentiments. A governmental body with the authority to use violence, yet void of Godly people is a sure way to remove his hand from a nation.

Violent campaigns were sanctioned and commanded by God. Scripture shows us that violence is a tool for God's justice. Ezekiel 25:17 is a sobering passage for pacifists to consider. It reads, "I will execute great vengeance on them with wrathful rebukes. Then they will know that I am the LORD, when I lay my vengeance upon them." Even the book of Revelation reveals judgment at the hand of Christ in many passages. Whether that violence is credited to God, the angels, Jesus after the second coming, or his people, we can confidently affirm that God does not need his image protected.

Justice is an indicator of God's disapproval of sin. Avoiding warfare is always the most desirable option, but not always the practical option. Retaliation after 9/11 is a debatable issue. In principle, the unwillingness to defend oneself or another is a refusal to protect God's image and carries a greater moral imperative. The extent and effects of war are often the friction points in cultural attitudes toward violence. I do not consider the opposition to enter a full-scale conflict unbiblical when considering the unnecessary loss of life and economic impact. I do oppose showing weakness to an enemy.

The true enemies we face are spiritual in nature and may manifest as a combatant on foreign soil, but most often strike from the same pew. Division in the church is spiritual warfare. When the enemy attacks physically it has done even larger damage spiritually. Non-violence, although commendable, becomes a slippery slope when it is the default answer to all problems. The issue of violence is an egregious vice concerning the welfare of civil servants when pacifism is weaponized to morally shun them. The truth is, physical violence is authorized with reservations—we must not become that which must be destroyed. The constraints and authority of civil servants are outlined in Romans 13.

Honorable men and women who defend others at the cost of their own lives abide in the love defined in John 15:13. Treating them as enemies of God leads to division and the deepening of traumatic wounds. This is exactly what the true enemy wants. The spiritual enemy always tries to take us out of the fight spiritually before resorting to physical warfare. He is not always strapped with explosives but the sentiments of emotional tyranny. Scripture tells us to take every thought captive to resist anything that goes to war against the knowledge of God. Bloodshed eventually ends, but for many the war in the mind never will.

PSYCHOLOGICAL WOUNDS

Even extreme grief may ultimately vent itself in violence, but more generally takes the form of apathy.

– Joseph Conrad

I remember sitting behind our hooch in Ramadi, playing guitar next to the words, "If you want peace, prepare for war" spray painted on the back wall. I understood what it meant, as did the rest of the Marines at Hurricane Point. But at home, many will never come to reconcile the relationship between violence and peace. Everyone wants peace. Activists advocate for it. Protestors demand it. But men of action are the ones who provide it. I signed my contract to defend the peace that was violated when four planes were taken hostage in September of 2001. I was handed a rifle, stripped of my identity, and placed in an environment where I might kill or be killed. I knew what I was getting myself into, but I didn't comprehend the full cost until it was over.

Marines run toward the sound of gunfire. "Locate, close with, and destroy the enemy" is the mantra of the Infantry. We are conditioned to confront the enemy by force. Yet, those who carry out God's justice are not immune to the physical, spiritual, and psychological consequences. There is a burden of individual recovery involved. Combatting a physical enemy as a civil servant involves a significant amount of discipline. The enemy at home is illusive and deceitful–waging the longest war in the mind.

The War on Terror exposed many servicemembers to traumatic conflict that hadn't been felt at a national level since Vietnam. Desert Storm was also a fierce fight, but it didn't have the volatility of economic, financial, and traumatic longevity to the same level that years of conflict accrues. The physiological

257

and psychological effects of OIF and OEF exponentially increased the prevalence of PTSD in society. Transitioning out of the military meant transitioning from a physical enemy to a psychological one. Additionally, the enemy becomes personal. A war between nations inevitably becomes a war for the soul.

American journalist, author, and filmmaker Sebastian Junger suggests that social factors play a significant role in veterans' trauma recovery. In his book *Tribe,* Junger details the downfalls of society in supporting homecoming war veterans. The loss of community is a significant contributor in how well veterans cope with traumatic exposure. Fighting a physical enemy is a mutual effort. The war continues to rage outside the military on a deeply personal level. Even with the support of others the psychological war is exclusively waged against the individual. Invisible wounds become the enemy at home. Strong communal support from family, friends, and professional services significantly contribute to a veteran's psychological battle. Sadly, the only person who can win the war of the mind is the individual. No amount of social support can conquer the spiritual enemy inside if the individual does not fight.

The psychological effects of PTSD become the battleground for the spiritual enemy. Shame and guilt become weapons of war. Isolation and fear tactics position the enemy for a spiritual kill. The spirit often dies before the body. An individual loses the psychological battle when they lose hope. Self-destruction and suicide are symptoms of a dying spirit. The spiritual enemy first seeks the soul, then attempts to manifest physically to take others. The loss of a veteran at home inevitably becomes collateral damage. Invisible wounds may not manifest outwardly until it is too late. When communal support is compromised psychological trauma becomes malignant.

Political ideologies have polarized the nation. Moral

justification for action was called into question when the United States invaded Iraq. People have asked me over the years how I felt about my time in Iraq. Some were more explicit in their questions than others, but the principle of their concern was the same: How did I feel about fighting an unjustified war? There is almost always an inquiry regarding my political posture toward the war, more specifically toward violence.

Politics were never a deciding factor in my commitment to the infantry. Someone was going to have to fight, and I wasn't going to wait to be called. Fighting was inevitable, but that did not mean I inherited the moral responsibility for the volatility of that fighting. The decision was out of my hands to choose which theatre I served in. I owned my decision to join unapologetically. What struck me about their questions were the subtle hints of shame. Some people have been overtly vocal about their opposition while the majority I've talked to over the years were simply curious. I always shared my experiences in Iraq from an apolitical perspective, and people generally received them well. Regardless of their stance I still felt the subtle pull of shame from the spiritual enemy.

I was thrust into a war of cultures. The world was not the same place at home it was when I graduated high school. The natural course of action would have been to bury my experiences and carry on. But I fought to avoid the internal decay that it would cause. I was open and honest with anyone who asked about my service. I answered their questions and didn't hold back. I could see horror in some faces, signaling regret in their inquiry. I answered their questions anyway, attempting to understand what they saw in me. It was the only way to truly gauge who I'd become in the eyes of society.

Guilt and shame are prevalent in PTSD. Guilt conjures a sense of responsibility for a traumatic event while shame attacks the individual's character. Many veterans feel personally responsible for the death or harm of a fellow servicemember, even if they were not at fault. Shame is the work of a spiritual foe. Shame tells someone they are the problem. American anti-war ideologies have significantly contributed to the internal shame veterans experience. Though they are not responsible for the political decision to go to war, veterans have been convicted of a moral crime for participation in it. Intentional or not, cultural values affect the post-traumatic growth process.

When guilt becomes shame, the mind shifts from incident to identity. Our sense of identity is affected by social factors. We learn who we are through the people we are associated with. If we are not careful, the world and all its spiritual forces at war against us will define us by everything we are not. The protests of Vietnam and the War on Terror can have a negative effect if they are directed toward veterans. The Dutch priest and professor Henri Nouwen wrote about how we have shifted from a guilt-based culture into a shame-based culture because we look to our peers for validation rather than our fathers. We are becoming more defined by the opinions of others rather than the authoritative words of our Father. Combatting shame becomes a matter of seeking our identity in Christ.

Pacifist ideals do little to assist the post-traumatic growth process. Passivity never emulates the full character of Christ. Most people are naturally conflict avoidant. Remaining passive in a world filled with violence is not the standard Jesus is calling us to. Conflict is inevitable. Sitting by when people are losing their lives is an act of fear not love. Returning home to be lectured about the morality of serving in the military or being told to remain silent about my experiences is a recipe for disaster.

Becoming a Marine was a pivotal part of my journey. Burying traumatic experiences to avoid offending sensibilities is spiritual suicide.

Apathy is another weapon the spiritual enemy uses to kill. Apathy destroys the willingness to pursue life with a purposeful heart. Everything seems out of reach and unworthy of effort. Apathy is nihilistic. King Solomon understood the weight of apathy when he uses the word meaningless 38 times in the book of Ecclesiastes. My first deployment journal from February 17, 2006 displays some of my first signs of apathy:

> "Music is the only thing giving me peace. Just one song can bring me back. Restore me. My mental state is completely different. I sit on Post looking through the iron sights of an M240 or my ACOG on my M-16 at the Iraqi's and don't hesitate one second to pull the trigger. My finger is just ounces of pressure away from putting a round through someone's chest. Men, children, I don't feel anything. I just wait for them to do something hostile. I try to justify what they're doing as hostile just so I can shoot. I'm fed up with these people, this country."

It's hard to process that these are my own thoughts at 19 years old. The emotional currency captured here gives insight into the depth of pain that has faded over time. The things I said at that stage in my life are a result of emotional collapse. I remember the feelings of frustration and hurt that hibernate behind those words. I remember the fatigue that grew over months of prolonged anxiety. I became indifferent to the threat of IEDs, small arms fire, and life itself. Death was just an escape from the daily struggle. Apathy suspended my emotional decay, adding up like a tab to be paid back home. Corporal Conley and Lieutenant Fitzgerald were killed by an IED the following day after that entry. My heart became calloused.

Apathy was my biggest struggle after service. The plethora of wounds were mine to reconcile and the tab was finally being paid. Tremors even surfaced when thinking about Ramadi. My separation from the tribe tore the void inside me wider. The relief that coming home was supposed to bring betrayed me. I was failing to adapt in the civilian world. I needed a community but felt disconnected from everyone. The psychological fight overwhelmed me. Apathy fractured my pillars. I sought refuge in the past out of fear of the future ahead of me. Nostalgia offered temporary relief in difficult times, but the constant pursuit of past pleasures left me empty.

I felt like a spectator to my own life after the military. I lost the community that embraced aggressive tendencies. The new community I gained was hesitant to accept me. Their political aversions to the war created a social cavity. People expected my old self, not the person I had become. I became frustrated and cynical, losing sight of the meaning of my sacrifice. I fought to retain the person everyone knew and loved. I hid in the past, refusing to step into the future. I spent years searching for the joy I had before all of the painful experiences. No matter how hard I tried, I could never find it. I was afraid of what God was calling me into.

Suppressing the infantry culture in my behavior was an unforeseen challenge. It became part of who I was. Aggression was a therapeutic release. Harsh language and intense comments had to be filtered in social settings. My ability to handle conflict grew weak in a culture driven by emotional sentiment. I wrote off my apathetic posture as persistence of a strong will. I didn't know how to communicate my stress without offending someone. Dark humor was a mechanism to deal with extreme trauma but was often received by other people as a warning of violent potential instead of a subtle cry for help.

FRACTURED PILLARS

Be sober-minded; be watchful. Your adversary the devil prowls around like a roaring lion, seeking someone to devour.

— 1 Peter 5:8

Scripture informs us about spiritual warfare. Matthew 11:12 says, "From the days of John the Baptist until now, the kingdom of heaven has been suffering violence, and the violent have been seizing it by force." Jesus' words here refer to the persecution he faced during his time on earth. The forces of darkness wage war against all people who represent Christ. Jesus also says in John 15:20, "If they persecuted me, they will also persecute you." Christians are under attack because we bear the image and likeness of God. Spiritual warfare intensifies as we grow closer in our relationship with God. Without action the enemy begins to rule our lives. Complacency is spiritual pacifism. Passivity in the spiritual realm is definitive, death is certain without action.

Our spiritual enemy has chosen to fight. He will do everything in his power to destroy us. John 10:10 says the thief comes to steal and kill and destroy. Physical violence is the ultimate goal of his destruction. The path to achieving that is by steadily tearing down the internal core of our being; hacking away at the cornerstone of our pillars. The safety and freedom we enjoy in America shelters us to the overt physical warfare that wages around the globe. Our efficiency in combatting a physical enemy leaves us extremely vulnerable to the spiritual one. If the enemy cannot kill us in body, then he will strike out our faith.

We entered a war when we deviated from God's commands. The Fall in Genesis chapter 3 captures the story of how the enemy strikes at our foundational pillars. In verse 4 the serpent tells the woman she can be like God. The enemy created doubt by skewing the sacred words of God, causing the woman to imagine herself as her own god. Verse 6 says the woman offers fruit to her husband, whereby sin becomes communal. The man allows this to happen rather than obeying God's command. In one strike the enemy was able to fracture humanity's image, community, and design. He lies to us about who we are. He stirs up division between us. He misguides our steps. He doesn't need to violently remove us from the planet, he just needs us to stop listening to God.

Complacency is a perversion of order. The spiritual enemy uses distractions to confuse us. Guilt, shame, and apathy are all forms of psychological warfare which separate us from God. We do the enemy's work if we fail to take ownership of our lives and our trauma. The adversary doesn't need to destroy us if he can cause us to destroy ourselves. If we are not careful to listen and obey the Word of God we will become isolated and rudderless with no sense of who we are. Luke 21:36 cautions, "But stay awake at all times, praying that you may have strength to escape all these things that are going to take place, and to stand before the Son of Man."

The Marines I served with have all been negatively affected by their time in service. Divorce, alcoholism, gambling, loss of limbs, PTSD, financial bankruptcy, diminished health, and the list goes on. Many were affected by poor decisions along the way. Some were victims to the decisions of others; returning home to find that their wives had an affair, depleted their bank accounts, lost custody of their children, or lost a family member. Their decision to serve did not exempt them from personal hardship. Pursuing justice in the wake of 9/11 was not without its challenges. Yet, given the chance to do it over again nearly every Marine I know would unequivocally say yes.

We can either be on the side of justice or on the side that perverts it. We have the choice of what to endure and those actions affect others. There is a consequence for sin and persecution for following Christ. Trouble will come either way. The enemy does not loiter when we choose justice. He brings the fight to us, striking at our foundation. The enemy's damage is focused on the strength of our foundational pillars. My spiritual battlefield assessment is not a matter of how well I am following the rules but in how strong my pillars are becoming. How much is my life reflecting Christ? What impact do I have on my community and how much am I growing in association with them? Am I serving God's kingdom in what I do each day?

Service in a combat theatre was a conscious choice. I was well informed on the physical battle ahead of me. Though I was unprepared for social isolation, anxiety, depression, shame, guilt, apathy, and every other symptom present in PTSD. These symptoms have become my cross to bear. I can shift blame anywhere I want but resolution begins by taking ownership. Trauma can easily separate us from God when we focus on our pain. As long as the enemy can keep our eyes on symptoms and not the problem he will slowly consume us with lies and deceit. Physical wholeness and healing are unattainable if we are disconnected from the Holy Spirit. King Solomon was given the full manifestation of physical prosperity, yet bled over the meaninglessness of it all. Our external well-being reflects our internal health.

Sacrifice will always cost us something. No good deed goes unpunished as the saying goes. Spiritual complacency begins with separation from God. We cannot abandon the work God is doing in the midst of our pain. Paul says, "Not only that, but we rejoice in our sufferings, knowing that suffering produces endurance, and endurance produces character, and character produces hope." (Romans 5:3-4). Look to the hope we have in Jesus who exemplified endurance through suffering. The modern Marine Corps is an all-volunteer force. We were not drafted, we willingly chose to fight knowing it could cost our lives. We are blessed to be dealing with the aftermath though it costs us something. We sacrificed so that others could live.

THE POWER OF THE CROSS

Love bears all things, believes all things, hopes all things, endures all things.

– 1 Corinthians 13:7

The Vietnam War was the first war where American servicemembers were rejected by fellow Americans on a major scale. Returning to an unsupportive community fueled their traumatic experience. Many failed to rationalize their experiences and struggled with the meaning of their service in a politically divided nation. Vietnam veterans buried their experiences and carried on with their lives pretending to be normal. In contrast, World War II veterans returned home to grand scale parades and over-whelming support on a national scale. The generation that survived the Depression and conquered a global evil were welcomed as heroes. Those who struggled to find meaning in their suffering had the support from their local communities which contributed to a massive population and economic boom.

The wars America has fought since WWII have not seen the same level of social support. The public's lack of support becomes a vice for veterans when it fractures social relationships. This type of adversity can be overcome when both parties come together. Veterans need a place to share their experiences and be understood while the public is able to learn vicariously the cost of sacrifice. Many veterans have lost their battle with PTSD and combat related trauma from isolation. We end up with an epidemic of suicide and a crucified Christ when the voice of public opinion echoes louder than the healing words of Christ.

If there is any hope for us in dealing with trauma it is modeled in the suffering of Christ on the cross. The military was a place where I felt connected to God despite the darkness of the world around me. As I watch the lives of other veterans who served I can't help but notice a trend. Many service members with visible symptoms of post-traumatic stress are separated from their respective branches. It's not that PTSD only affects veterans, but that many continue to work through their trauma during their periods of service. It's when their purpose is lost and there is no community to replace the military that their lives often deteriorate. Many reenlist for this very reason.

One intersection of our struggle with trauma is between adversity and purpose. Without a clear sense of direction or meaning we lose our will to keep moving. Without purpose adversity consumes us. We become helpless to fight the hard battles and begin seeking out temporary fulfillment to get by. We pacify our pain with distractions like alcohol, drugs, sex, and dopamine fixes. These things can never sustain us. We place burdens on loved ones that they were not meant to carry. All people, especially people with deep voids and immeasurable pain, are left hopeless without a Savior. We must seek the one who satiates our eternal desires.

Adversity is not a burden that justifies giving up. It is a call to bring forth the change that must take place. Adversity advocates action. I can't help but see defeat in those who reject justified violence on the basis of its vulgarity. Sometimes violence is senseless, but other times it is necessary. We will never distinguish what is necessary and what is senseless if we live with the illusion of a perfect world manifested by anything other than God. Adversity is the enemy attempting to separate us from Christ. It is the testing of our faith.

People often look astonished when I share my Iraq experiences. They've come to know the person I am outside the military and find it puzzling to hear about what I've been through. More than a few times I've heard the phrase, "how are you normal?" My response is always, "define normal." It's a strange compliment to be called normal but also misdirecting. If they could see the invisible pain, then they would see that normality is subjective. Those close to me have seen the external effects of my internal world. In reality, this question illustrates the present state of our culture that has long been removed from true suffering.

In relation to violence, those serving in the armed forces benefit from a compartmentalized experience. Wars are fought in foreign countries with significant periods of travel to get there and back. Comparatively, we live in a period of history with minimal human suffering. In the United States especially, we are far removed from the intimate scarring of hostile conflict. The freedom and peace we enjoy becomes the vice from which veterans suffer as outsiders. The blessing becomes the curse.

Still, violence is not completely outsourced. First responders fight our battles at home. Consider societal violence in contrast to overt warfare. Law enforcers, fire fighters, and medical personnel not only experience the same grotesque death of our brothers and sisters but also with the families of their own communities. Additionally, they don't have the luxury of a compartmentalized war zone buffered with a period of travel before "reintegrating" into society. They go from death to the dinner table.

I am blessed to have a father who served a successful career as a police officer. I can't point to every facet of how his resiliency was passed on to me, but I can see the effects of it when I hear people compliment the way I carry myself in society. One random afternoon I was walking through an area of my hometown with my father when he recalled a story of a call he responded to in the exact place we were standing. A woman had been brutally murdered and left unidentifiable without forensics. The deceased was someone my father had gone to school with growing up. As he told me this I had the very same thought that everyone had of me. *How are you normal?* How had I never picked up on any symptoms of traumatic stress growing up?

He told me this story a few years after the military, knowing I had already seen my share of human destruction. I knew he experienced similar violence, but hearing it in that moment meant he understood the impact of evil in the world. Yet, he carried on like it was just another day in the office. I immediately understood the effect of fighting a compartmentalized war in a far-off country. Ultimately, the inextinguishable hope I have for life came through the experiences of death and suffering. I don't fear the experience of violence when fought next to my brothers in a collective effort. I fear that one day a war will come to our land and we will be grossly unprepared to face it on a social and psychological level.

The words of my Company Commander spoken prior to my first deployment still ring true, "We are in the business of killing and business is good." Many see the image of a rifle and think only of the pervasiveness of death. Those who have carried one know how life can be sustained or destroyed by the will of the operator. We have been given a stewardship to preserve life. When we live with a debilitating traumatic experience, our own lives are the ones in need of preservation. The unresolved destruction of the external world points to the collapse of an internal one. Self-destruction is the natural result. Suicide becomes the permanent solution for a reconcilable grief. The tragedy is in not recognizing a hope that has always, and will always, be available to us.

Could we ever truly understand God's love without suffering in the world? Even if we take an atheistic approach to life, we aren't excused from pain. Christ offers us hope, not an ideology. He offers us a way because he *is* the way. I see my father as resilient because he always carried his cross with the understanding that death is not the end but the beginning of life. The adversary wins when we lose sight of our reason for living. 1 Peter 3:15a says, "but in your hearts honor Christ the Lord as holy, always being prepared to make a defense to anyone who asks you for a reason for the hope that is in you." It is a blessing to be able to suffer for Christ in this life.

Crucifixion is a violent persecution. Christ served others with an inextinguishable love and was murdered for it. He was cursed for our benefit. Christ reclaimed the image of the cross from a cultural symbol of death and torture to become an international marker of love and healing. We are called to carry our own crosses daily. Those who combat violence share in the suffering of Christ. Our recovery through trauma is not just for the sake of normalcy, it serves a source of hope for someone else. The extent of our suffering is proportional to the level of character required for the calling. Marines may be in the business of killing, but Jesus is in the business of redemption. Both work for the glory and justice of God who is our source of life.

CHAPTER SEVENTEEN
WOUNDED HEALER

I heard a light sigh, and then my heart stood still, stopped dead short by an exulting and terrible cry, by the cry of inconceivable triumph and of unspeakable pain.

    – Joseph Conrad

I'M NOT A FATHER, but I look forward to having a daughter someday. I have no sisters and my extended family is relatively small with no aunts. The thought of raising girls is exciting despite all the warnings I hear from fathers of daughters. There's something about the innocence of a child and the responsibility of influence that makes life more meaningful. I'm not opposed to having a son. I just have a stronger desire to look upon the purest version of my wife rather than myself, especially coming from a predominantly male family. God gives us our desires and this one has stayed with me for a while.

That being said, I believe the most pain a person can experience is the loss of a child. I can't relate to this grief, but I greatly sympathize with those who have. One day this will be a very real fear for me. Having children is a challenging yet wonderful experience. That is why losing a child can be the most unimaginable pain a person can experience. It has destroyed marriages, led to substance abuse, self-harm, even additional loss of life. However, pain is not meaningless. No matter what circumstance take place, the living must keep the memory of our loved ones alive.

I remember stepping off the bus in Twenty-Nine Palms when my unit returned from our first deployment. Seeing all the families on the field anxiously awaiting our arrival was the fulfillment of a moment we all knew we might not see. Two moments will stay with me forever from that night. The first was seeing Swanberg's mother who came to see everyone else's son return despite losing her own. She confronted every reminder of her son in order to share in everyone else's joy. The second was seeing my own mother who upon seeing me said that my appearance had changed. After only eight months of being away, she could see how the world had taken something from me and collaterally from her.

Most people avoid things that remind them of a traumatic experience. This can be places they've shared with a loved one, music, movies, colognes or perfumes, and even certain com-munities. Many are never able to accept the loss of a loved one. Acceptance doesn't mean forgetting and moving on but learning how to live with a debilitating wound. The way in which we do this is by keeping the memory of a person alive. We hold memorial events, dedicate plaques, name streets, buildings, and ships after people. We write laws to honor the memory of someone in order to prevent further tragedy for others. There is a unique component to our design that separates us from the rest of the animal kingdom: how we honor our dead. This isn't restricted to the battlefield either. It is a universal process of healing. It is an innate human act not to give death the final word.

RESILIENCE

> He *himself was a living proof that the fieriest energy is not*
> *incompatible with the ability to relax.*
>
> — Marcus Aurelius

An entry from my first deployment journal captures a moment when I was overcome with pain. February 28th, 2006:

"This entire week 3/7 has been getting hit all over. Kilo Co. lost more Marines to IEDs. HP took a mortar at the front gate, 2 were hit, 1 was killed. The phones and internet have been down for about a week. I finally got to call my family and tell them I'm OK. I've been talking to friends here and found out a lot about different things. A week out of SOI last year, 4 guys I knew were sent straight to Iraq. They were killed in a helicopter crash. A few more from 2/7 in Fallujah were killed… I find it hard to believe sometimes, how I'm still alive after all this."

It is hard to read my own apathetic words from when I was nineteen-years-old. I still haven't fully processed those experiences, but I have found meaning in them. We don't have to understand what we are going through in order for it to serve a purpose. Faith requires that we find the courage to continue in our suffering knowing that one day we will be reconciled to God. We are at the starting point of our faith when things don't make sense. Suffering is an opportunity for growth. I struggle to find meaning in the midst of hard seasons. Yet, every time I've reflected back on trying times I've learned something and seen growth in areas I was weak. More importantly, I've seen how my life has affected those around me.

Faith is trusting that God's purpose is being fulfilled

regardless of our pain. The fear of suffering is a natural human defense mechanism. Despite our internal warning systems people will endure immense suffering for what they love, including losing their own lives. The reality is that people are more afraid of meaninglessness than they are of pain. When we don't know why we are suffering our pain has no purpose. Faith grows through suffering because we are moving closer to Christ. The measure of a mature faith is how consistent we love while enduring suffering. Without Christ, life has no value.

The most prominent story of suffering next to Christ is that of Job. Job is considered the oldest book in the Bible (estimated 2100-1800 B.C.), making the story of human suffering a central experience in life. Job lost his wealth, his family, and his health yet in all this he remained faithful to God. The main point of Job's story is not to demonstrate humanity's capacity for pain (which it does), but to recognize God's faithfulness and sovereignty while suffering. Job's trials also point us to a posture of humility. Job was not an emotionless stoic who gracefully accepted the hardships he endured. His complaints demonstrated a self-righteous attitude toward suffering, which God himself corrected. The moment Job embraced his suffering and admitted his lack of understanding, God blessed him with double of what he had previously. Job remained faithful to God, not knowing it would result in a major blessing.

Christianity as a formal religion began roughly two-thousand years after Job. The most notorious missionary of the gospel was the Apostle Paul. Paul experienced great suffering on his missionary journeys as well. His suffering marks an interesting parallel to Job's because he understood why he was suffering and willingly participated. Compared to Paul, Job was a bystander. Paul's suffering was a result of his calling and eventually costed him his life. God was gracious to bless Job with twice what he had in his lifetime whereas Paul faced death with certainty. Neither Job nor the Apostle Paul were faithful because of any perceived temporal blessing in life. Both were faithful to God because they recognized his sovereignty and that he was the greatest treasure of all.

There is a well-known Hymn called "It Is Well With My Soul" written by a man named Horatio Spafford in the late 19th century. Spafford wrote the song at the threshold of his pain to recognize God's peace during trials. In 1871, Spafford faced financial ruin due to the Great Chicago Fire. He later lost his four daughters in a shipwreck where 226 people lost their lives on a transatlantic voyage to Europe. His wife Anna survived the wreck and continued to England, sending a telegram about the accident. On his journey to meet Anna in England, Spafford wrote the famous hymn over the waters where his daughters had perished. A fellow survivor of the wreck recounts the words of Anna saying, "God gave me four daughters. Now they have been taken from me. Someday I will understand why."

Viktor Frankl was a Holocaust survivor who tells of the Nazi death camps in his book *Man's Search for Meaning*. He covers grueling details of the prisoners' struggle to survive under barbaric conditions. Frankl lost his parents, brother, and wife in the concentration camps. He admits that some prisoners, who became trustees within the camp system, sought favor with the guards by denying the humanity of their own group for the sake of personal security. They were more brutal to fellow prisoners than the guards themselves. The violence of the Nazi regime turned victims into perpetrators. Still, Frankl emerged with the belief that God alone should be feared above all human evil.

Frankl talks about the disillusionment and bitterness of his fellow prisoners after liberation from the Nazi concentration camp that resulted in a moral decay. Most prisoners lost their will to live then were suddenly granted a freedom they did not expect. The damage of tremendous suffering compromised the character of many. Frankl however, found meaning in his suffering. He had every right to lash out against the world after what he endured. Instead he chose to bring hope to the lives of others despite the loss of his family and his wife. He continued his work as a Psychologist and pioneered a form of existential therapy so others could find healing in their own suffering.

Frankl and Spafford did not have a supernatural experience with God the way Job and Paul did. Pain and suffering are not alleviated exclusively through divine interaction. Scripture reveals the character of Christ who has given us his Holy Spirit. Peace comes only through Christ, and we have been given this gift through his Word. These examples show us that we will suffer for the actions we take as well as the actions we do not take. The question is not *if* we will experience difficulties but *when*. And when we do, are we going to suffer for Christ or without him? Suffering is the great conduit for meaning.

The enemy takes ground every time we surrender. Suffering obligates us to choose between peace or destruction. Able ruled over the enemy; Cain let it govern him. Jesus promises a peace that surpasses all understanding when we place our faith in him. Pain causes our faith to grow. Paul writes in Romans 8:18, "For I consider that the sufferings of this present time are not worth comparing with the glory that is to be revealed to us." Meaning is what makes suffering worth enduring. Christ endured suffering to liberate the human spirit. Galatians says it is for freedom that we have been set free. The Apostle Paul brought the message of this freedom to the rest of the Gentile world by writing letters to various Christian churches from prison. No depth of suffering can overcome the power of Christ. The common thread that links the unconquerable spirits of Job and Paul; of Spafford and Frankl; is an understanding that peace is found in the eternal and is worth losing everything for.

THE PERSON OF CHRIST

> *For to us a child is born, to us a son is given; and the government shall be upon his shoulder, and his name shall be called Wonderful Counselor, Mighty God, Everlasting Father, Prince of Peace.*
>
> — Isaiah 9:6

In Matthew 4:1-11, Jesus is led into the wilderness to be tempted by the devil. Similarly, Job was allowed to be tempted by the devil (Job 1:8). Jesus fasted for 40 days when he was confronted by the devil in his hunger. Then he was tempted to prove himself as the Son of God by invoking God's protection over him. Finally, he was tempted with having dominion over all the kingdoms on earth. His temporal needs, identity, and authority were attacked in an attempt to derail him from his purpose. Every one of his responses proves to us his divine nature and legitimacy. Jesus never compromised his purpose in exchange for temporal relief or illegitimate power.

The devil is the father of lies. Jesus' restraint through physical and metaphysical distractions is a testament to his unity with God. His triumph over temptation is a triumph for mankind. Jesus displayed two powerful characteristics through his death and suffering. The first was his willingness to die as a substitute for an undeserving humanity. Jesus took on the wrath of God in our place. The second was his obedience throughout his suffering. Jesus exemplified love through the worst pain of human experience. He endured unimaginable trauma, betrayal, humiliation, and isolation all for our benefit. He was cursed for our sin and bore the consequence we should have received, and he did it freely. By raising himself from the dead, he proves the he is the Son of God and declares our freedom.

In regard to Christ's emotional suffering, the feeding of 5,000 people with five loaves of bread and two fish in Matthew chapter 14 is an extraordinary account of his emotional resolve when he meets the needs of others despite his own weakness. The event is more than miraculously meeting their nutritional needs. The context from which Jesus approaches this encounter is what highlights his character. Chapter 14 actually begins with the story of John the Baptist's death. Matthew 14:10-12 says,

> "He sent and had John beheaded in the prison, and his head was brought on a platter and given to the girl, and she brought it to her mother. And his disciples came and took the body and buried it, and they went and told Jesus."

This gruesome murder carried out on a defenseless prisoner is much like the executions of ISIS today. This kind of violence can move us beyond sorrow to stir up vengeance in the heart. The amount of grief Jesus felt in that moment drove him to retreat to a remote place to be alone. The crowds relentlessly pursued him without regard for his own sorrow. It would have been understandable for Jesus to have turned them away until he was of sound mind to receive their concerns. But his love for them was more important than his own well-being. Matthew 14:14 states, "When he went ashore he saw a great crowd, and he had compassion on them and healed their sick." It was that evening he fed the 5,000. Death cannot burden God's love.

Through temptation and suffering Jesus never deviated

from accomplishing God's will. He was obedient to the point of death even suffering in every human capacity. He remained faithful though his disciples abandoned him. He remained faithful though he was betrayed by the crowds who shouted "Hosanna" on Sunday then chanted "crucify him" on Friday. He remained faithful through torture and public humiliation on the cross. But the most difficult trial he endured was in the calm of the Garden of Gethsemane. Matthew 26 says his soul was sorrowful to the point of death and he sweat blood in anticipation of fulfilling his purpose. The last chance the devil had to derail him was in the temptation to avoid suffering on the cross.

The will of God is more important than comfort. It was complete obedience that sustained Jesus in his earthly ministry. He withstood more than any human, never ceasing to serve others in his sorrow and pain. There was a purpose in everything he endured. He even calls Peter "Satan" and Judas "Friend" in relation to God's will. Obedience to God's will enables us to endure trials and temptations even when our suffering feels meaningless. It is by his wounds that we are healed; that we are able to endure our own suffering; that we become children of God.

RECONSTRUCTING PILLARS

But, after the fires and the wrath, but, after searching and pain, His mercy opens us a path to live with ourselves again.
                                                    – Rudyard Kipling

It is common for someone who loses a child to also lose their faith. Violence and tragedy show us the darkest potential of humanity. Children show us the greatest potential. It is the greatest joy to be called children of God. We are his greatest work in the world. We radiate his image and power in what we say and do. We have a great capacity to love as well as a treacherous capacity to harm. While we reflect God, our children reflect us. To lose them is to lose a piece of ourselves. Equally, to lose a child is to lose our sense of God's presence.

Mourning is something I overlooked when my own life was on the line. My reverence for the burial process was reserved for everyone else but myself. My state of mind remained in survival mode for years after I got out of the military. My own death was something I refused to process. I was apathetic toward my own life when the threat of losing it was imminent. That apathy remained until I saw the value of my life from a different perspective. I was having a conversation with my parents roughly six years after the Marine Corps while we were out to dinner one night. We were talking about my grandmother who passed away and the topic of funeral services came up as did my calloused opinions toward death.

I hadn't given much thought to the grieving process. I had been to more funerals than weddings at the time. Death had become so common that I was desensitized in my ability to process it. I understood the need for closure and to feel the nearness of a loved one. I saw the grieving process differently for myself because the thought of being mangled in an explosion and having a closed casket funeral made me want to avoid having others in my physical proximity after I'm gone. I wanted whatever was left to be ashes and sent back into the dust of the earth. I wanted people to move on and enjoy life with me as a memory. I didn't want to be restricted to a gravesite.

"I just don't get why funerals are so expensive," I said.

"Funerals are a large business and people pay a lot of money to have closure for family members," my mother responded.

"I know, but it's a scam to charge thousands of dollars for

280

an extravagant box that ends up buried in the ground. Put me in a cheap wood box and dig a hole somewhere important to the people that are still alive, or scatter my ashes somewhere meaningful. I don't want to be a financial burden on people after I'm gone. It feels unethical to take advantage of people's emotional grief for large profit."

She hesitated before responding, "When you were overseas we arranged plans with a local cemetery if something had happened to you."

Death was a very clear reality for my parents. The costs involved meant nothing as long as my memory was kept alive for them. I trespassed all over their method of grieving not realizing how it diminished the value of my own life. I needed my parent's signature when I signed my contract because I was underage. They kept asking if I was sure this was what I wanted to do, implying; *Do you know what you're getting yourself into? Do you know what this could cost you, what this could cost us?* It had nothing to do with the financial burden. They authored their own guilt by sending me into the unknown. Into a world completely outside of their control and with a sense of responsibility. I never considered the value of my life until that night. They allowed me to choose for myself beyond their own reservations but refused to show the same disregard for the memory of me. Sometimes we don't realize how valuable we are until someone else loves us.

Children have an innocence that humanity holds in the highest regard. They have intrinsic value that must be protected. When the world takes them from us it shatters our worldview. It violates the perceived rules of life. Without God these experiences will leave us permanently defeated. God is intimately familiar with our pain. He endlessly pursues us and shows us our value. As humanity destroyed itself with violence and buried itself in sin, God stepped into human history and freely offered his own Son for our benefit. He gave him up to die a terrifying death and took on the world's pain because he loves us.

Our physical bodies know the shutdown of emotional wounds. Wounds that pierce us on a spiritual level and alter our perception of reality. Our souls cry out for comfort that we often seek through destructive habits. We satiate our suffering with things but are still left with a God-sized hole in our hearts. We find other ways to fix the problems we can't rationalize. I became aggressive, desensitized, irrational, and apathetic. Few find hope in dire circumstances and make peace with un-answered questions. Even fewer can find the strength to serve others in the midst of their own suffering and go to war against the powers of darkness. The pain we go through will mean nothing if we allow it to destroy the work God is doing through us.

After the Marine Corps I lost sight of my own worth. I had four years of emotional and moral baggage to unpack. I chased relationships to fill the void of intimacy. I chased accomplishments to feel valued. I never touched alcohol as a way to cope but sank into depressive seasons of self-doubt and emptiness. I grew cynical, blaming everyone for the slightest infractions. My emotional state would go from zero to a hundred in a second. I had no direction or community. My pillars were fractured, and the world began to define what was valuable in my life.

Music helped to escape the meaninglessness of life. I forced myself to seek out community when waiting for it to come got me nowhere. In college, I connected with a small group of friends through similar interests in music. Our lives were on different paths but our common venue allowed us to support each other in our seasonal needs. Despite knowing they would never fill a major void in my life, I was content for the first time in years. I volunteered with a church after graduation to stay in community. Most people had different values and beliefs, but I had discovered meaning in what I was able to do for others. Every time I've retreated from life God found a way to bring me back into his presence. I rediscovered a love without condition as I served beyond my own needs.

We are God's image to the world; we have the responsibility of displaying the resilience of his character. God is shaping us for his purpose through our trials. Distractions will derail us from serving if we are only focused on ourselves. Trials are designed to strengthen us. We are a model of hope to others as we rise through the ashes of life. To defy the worlds expectations through faith is to point others in the direction of God. It is not by our power but the power of God that is in us. Through obedience and prayer, God will reconstruct broken pillars.

We become living proof that joy is attainable in any condition. I wish I could say I've used every experience to bring hope to others in their trials. I've seen the limitation of hope in my life and the tragedies I let grow because of my apathy. The overwhelming betrayal of a loved one weeks before my first deployment and the goodbyes to family and friends were enough to feel dead before facing a single physical threat. We are being made strong not for our own sake but for the sake of others. Our lives may be the only Bible someone reads.

Love is always a choice. It should not be reduced to a feeling or the suppression of pain. Love is compassion under the weight of sorrow. Love is the persistent commitment to the mission at hand. Jesus' purpose was to reconcile humanity to the created order, and he made it possible through his obedience to God's will. Jesus is the ultimate expression of love in a world plagued with suffering. To be loved by Christ is to be loved by

CALLOUSED HEART

someone who holds all the worlds pain and endlessly comforts us so we may never lose our way.

CHAPTER EIGHTEEN
CALLOUSED HEART

The awful thing is that beauty is mysterious as well as terrible.
God and the devil are fighting there and the battlefield is the
heart of man.

　　– Fyodor Dostoevsky

KENNETH RICK FIRST DEPLOYED to Ramadi, Iraq as a Lance Corporal and
SAW gunner with 3rd Battalion 7th Marines, Weapons Company
in April of 2007. He finished his senior year of high school a
year and a half before when I was in Ramadi on my first
deployment. We conducted humanitarian missions to help rebuild the war-torn provincial capital for a majority of the
deployment. We handed out rice and flour bags to local families,
oversaw the recruitment of a national police and military force,
contracted locals to rebuild structures, and cleared travel routes
as well as provided security patrols throughout local neighbor-
hoods. His first deployment and my first were quite different.

　　On many occasions, local Iraqi children followed military
Humvees, asking for handouts of candy and soccer balls. The
monotony of routine patrols in a simmering city eased tensions
for us and allowed us to appreciate their presence. Wearing full

combat gear in the scorching heat, somewhere north of 115 degrees, made for a near delusional mental state where kids could often acquire loose gear when Marines dismounted from our trucks.

Weapons Company held security in the Al Warar district near some run-down apartments one afternoon while handing out goodies to locals. Echo Section dismounted to survey the area. Lance Corporal Rick made his way into a nearby field with his M249 SAW. I saw him kneel to meet a small Iraqi girl face to face, and I happened to snap a picture of her fearless interaction. At the time, I didn't know how much it would capture the essence of the Marine Corps spirit in a single moment.

Five years later, the irony hit. I was now the one in school when I reached out to Rick to see how he was doing. He was deployed to Afghanistan on his fourth combat deployment. He was a Sergeant with 1st Battalion 7th Marine Regiment (1/7). I made plans to send out a few care packages and asked if he remembered the picture of him and the Iraqi girl. The company I worked for at the time had it framed and mounted with a few other military images in the office. I reached out to let him know how people were responding to it. In less than three weeks from our conversation, Sgt. Rick was engaged in a battle that earned him the nation's third highest medal for valor.

Marines from 1/7 "Dog" Company were inserted via helicopter heading west from Sangin for the village of Qaleh-Ye Gaz. They were tasked with a reconnaissance mission in support of special operations in a region that it is said to have never been taken, let alone maintained, in its history. The Marines from 1st Platoon arranged to remain in the valley for a week to survey the area. Early in the morning of their insert, their first recon-naissance patrol was ambushed by enemy small-arms and medium machine gun fire. Those still in the landing zone received incoming mortar and small-arms fire at the same time. Sergeant Rick and his squad pushed to cover from the landing zone while receiving reports of a wounded Marine from an adjacent platoon. Rick and his squad regrouped with only a few Afghan soldiers wounded. One with shrapnel to the face.

First Platoon began planning a counterattack to surprise the enemy when a call over the radio stated the Marine who was wounded in their adjacent platoon had passed from being shot in the throat. With three hours into their patrol and a deceiving lull in the fight, 1st Platoon's position was further tested when six grenades came flying over a wall of their compound. Sergeant Rick hastily shielded his SAW gunner and radio operator with his own body while pushing them to cover. Miraculously, only three men received shrapnel wounds. Sergeant Rick was hit in the butt cheek, while Caleb, 1st Squad's Corpsman, took a piece of shrapnel to the tricep. Both injuries were non-life-threatening. Rick and Caleb refused to report their wounds to avoid being pulled from the mission. Marc however, the platoon's radio operator, sustained a collapsed lung from a piece of searing metal. The severity of his wound required an immediate Cas-Evac.

A landing zone team consisting of five Marines and an engineer named Stevens was being led by 1st Squad's 1st team leader, Jeremy. Jeremy jumped at the chance to lead. The team set out to clear a landing zone (LZ) for a helo while Rick and another squad leader provided cover from the roof of their compound. Only five minutes after departing, the LZ team was hit by an RPG and PKM fire. Stevens was killed instantly and another was knocked unconscious behind him. Rick suppressed the enemy fire with his M203 grenade launcher in an attempt to cover Jeremy's teams' egress. In what seemed like only a second later, Jeremy ran into the patrol base, nearly out of breath telling everyone that Stevens was gone. Rick and the other squad leader who was suppressing fire with him both jumped off the roof and sprang into action without saying a word. Rick grabbed the M249 SAW while the other squad leader and Jeremy grabbed the stretcher. All three made the 200-yard dash out to Steven's body while exposed to enemy fire. Rick suppressed enemy fire with the SAW while the two other Marines evacuated Stevens. They were only five hours into the mission and sustained two KIA and one casualty awaiting evac.

The following day, Rick led a patrol toward a known enemy position where his platoon had been receiving contact in an effort to make the position inoperable. Niall, Steven's replacement, identified the location of an IED when the squad approached the enemy position. He marked the IED's location in accordance with their unit procedures before Rick cleared the compound single-handedly. After only finding a few spent AK casings and three RPG fins, he exited the courtyard to link up with his squad. Just then, the enemy began firing from deep down an alley attempting to lure the Marines in. Rick and his Marines moved to immediate cover.

In their hasty movement toward cover, Niall set off the IED he had marked fifteen minutes prior. Niall was blown into an irrigation canal full of septic water. With 1st Squad behind cover and with orders to hold their position, Rick made his way to Niall to begin treating his injuries. Niall lost one leg up to the hip bone and most of the other up to the thigh, he was seriously wounded by shrapnel in both of his arms and would undoubtedly be a quadruple amputee. Sadly, he succumbed to his wounds. With only two days into the mission, the total damage had resulted in 4 KIA and 13 American and Afghan casualties.

Two years later, Sergeant Rick was awarded the Silver Star

Medal for his actions in Afghanistan. Jeremy received the Navy Commendation Medal with Valor, which triggered a fight to be upgraded to the Silver Star. Had it not been for the courageous actions of the Marines who sacrificed their own lives, others might have been counted among the lost.

I heard about Dog Company's actions in Afghanistan a few years later when Rick's award was made public. It was the first time I felt the personal separation of conflict from the other side of the world. Pride and remorse swirled inside me while trying to comprehend the duality of two worlds. I felt closer to the Marines who carried on our legacy than the college students next to me. I grew callous toward everyday life and hungry to get back into the fight.

In Iraq, I fought Al-Qaeda insurgents on the streets of Ramadi when Rick was a senior in High School. We patrolled the same streets a few years later, conducting humanitarian efforts, handing out candy, and practicing Arabic with locals. He battled Taliban fighters in Afghanistan when I was starting my undergrad. The cycle of life and death aired alternate realities. The image of Rick with the Iraqi girl came to mind in a full display of esprit de corps. He knelt down to meet the innocent, eye to eye with the same weapon system that suppressed an enemy in another theater. Marines are masters of death and destruction. They are shameless in bloodshed and callous in their delivery. Through sacrifice, they have received a forged spirit which battles against the corruption of the world.

UNDERSTANDING CALLOUS

For this people's heart has grown dull, and with their ears
they can barely hear, and their eyes they have closed, lest they

should see with their eyes and hear with their ears and understand with their heart and turn, and I would heal them.

– Matthew 13:15

There are two primary descriptions for the use of the word Callous: (1) Being hardened and thickened, and (2) feeling or showing no sympathy for others…indifference to suffering. The word *Callous* is often used to describe a negative effect. The Diagnostic and Statistical Manual of Mental Disorders (DSM-5) used by psychologists defines the trait as, "Lack of empathy - disregards and is unconcerned about the feelings of others." Callousness is a symptom in most Cluster B personality disorders, which consist of disorders commonly associated with criminal behavior. The first definition of callous can be a positive benefit when dealing with physical resilience. The ability to work with tools or play an instrument is sustained by calloused hands. The second definition is a detriment in matters of spirituality.

The physical condition of hardened skin from extended physical work allows for an efficiency in performance. I have calloused fingertips on my left hand from playing guitar for many years. We also use this definition to describe emotional hardening such as having "thick skin" to symbolize being able to handle criticism. This could also be considered a calloused mind. Callouses are a positive attribute in this regard. They are a mark of resilience and strength. Playing guitar for hours without ceasing comes at the cost of sensitivity in feeling. Loss of sensitivity in this context is also a strength. Great musicianship outweighs the cost involved because the product is worth the sacrifice.

Biblically speaking, callousness is primarily a matter of spiritual health. 2 Timothy 3:2-3 contextualizes callousness in the brutality of humanity in the end times, "For people will be lovers of self, lovers of money, proud, arrogant, abusive, disobedient to their parents, ungrateful, unholy, *heartless*, unappeasable, slanderous, without self-control, brutal, not loving good." The word 'callous' even appears in this verse in the Amplified translation. Jesus warns about the callousing of our hearts because it is the rejection of God. A calloused heart leads to a godless and violent world. Hardening of the heart in this context results in contempt where brutality seeps out from within. Callousing in 2 Timothy falls within the second definition. This is in essence a spiritual callousness.

Similarly, emotional callousing is the metaphorical thickening of layers of the heart. The Bible makes numerous references to the hardening of people's hearts. The first mention is in the book of Exodus when Pharaoh hardens his heart toward God and does not allow the Israelites to depart from Egypt. Instances of spiritual callousness illustrate rebellion against God anytime a heart becomes hardened. The Apostle Paul uses the cultural imagery of circumcision of the heart (Romans 9:29) as a matter of metaphorically removing excess layers. Paul is talking about maintaining a pure heart by obeying God's commands and walking in his ways. Remember that James 1:27 adds that true religion is not only to take care of the orphans and widows but to remain unstained by the world. These verses shed light on the battle for the human heart and the function of its layers.

The Bible uses the metaphor of callousing in both definitions of the word. The world is at war with God and therefore has calloused hearts. That is why Jesus tells us the world will hate us because it first hated him. The enemy seeks to destroy us and anything resembling God. Jeremiah 17:9 says the human heart is deceitful and beyond cure. However, to walk

291

with Christ is to guard our hearts against the ways of the world. Proverbs 4:23 tells us to guard our hearts because everything we do flows from it. There is an apparent conflict in understanding the capacity of the human heart. Are we to guard what is most deceitful and corrupt? Or is there a duality in the way the bible contextualizes *callous*, linguistics aside.

To war against Christ is to harden our hearts against him through rebellion. To war against evil is to guard our hearts against the ways of the world and walk in obedience. Our hearts will be calloused through rebellion or obedience. We are caught in the middle of a spiritual war for the human heart. We don't have the option remove ourselves from the battle. The difference is in the calloused layers acting as an external or internal armor of the heart. Our hearts will be calloused either in the ability to withstand the enemy's attacks or by becoming indifferent toward God.

In warfare, the dichotomy of callousing can be polarizing. The human heart grows indifferent toward suffering when experiencing extreme fear and emotional strain. The process is known as desensitization. Desensitization is defined as "a treatment or process that diminishes emotional responsiveness to a negative, aversive, or positive stimulus after repeated exposure to it." Continual exposure to human suffering in war diminishes our emotional responses. The effects are often catastrophic. Many use their suffering to justify contempt for the world and God. Others become apathetic to any form of hope. This is precisely what Frankl captures after his liberation from the Nazi concentration camps. This type of callousness stains us and what Scripture warns us to guard against.

Despite the negative effects of emotional callousing, a certain amount of desensitization can actually become a strength.

There is a utility in emotional callousness but it comes at a much greater cost than extra layers of skin. Marine Corps leadership manuals explain the concept of "Mission Accomplishment", in which Marines are entrusted with carrying out their mission at all costs (this does not include ethical or moral violations). The natural human resistance toward violence is not a justified reason for failing to execute a mission. There is a reason why warriors fight wars and not poets.

Exposure to violence desensitized my emotional response toward human suffering. The threat of death or great bodily injury grew null from prolonged exposure. The emotional emptiness I felt after my first patrol prepared me to face the next seven months of extreme trauma. The lack of sensitivity I felt gave me the clarity of mind to process things like positive identification of threats, returning fire, and securing mass casualty sites. I was able to complete my tasks proficiently without emotional distractions. I grew callous toward human suffering in Ramadi for the immediate benefit of survival, the homeland benefit of peace, but at the future cost of my own well-being.

Emotional callousness is exponentially greater when playing an active role in someone else's suffering. Suppressing human sensibilities in the short-term has long-term benefits; even if those benefits involve living with remorse, shame, guilt, post-traumatic stress, depression, or any other symptom because each one of these problems is an alternative to death. How we respond to our traumatic experiences will ultimately determine the condition of our hearts. A calloused heart is not a hopeless one. Hope is possible so long as we remain obedient to God's Word. The Holy Spirit was able to break through the dead layers of my heart. My recovery after the military reflects the power God has to transform the human heart.

CALLOUSED HEART

Saul's transformation to Paul on the road to Damascus in Acts 9 is a miraculous account of God's ability to transform the heart of a man. Saul was someone who exemplified the definition of callous in the biblical context of rebellion toward God. His transformation to Paul by the Holy Spirit empowered him to steward the Christian faith despite the inevitable suffering involved. His obedience to Christ repurposed the layers of his heart. His fear of death diminished allowing him to bear the world's attacks with courage, even to the point of losing his life.

STRENGTH IN WEAKNESS

I sung of chaos and eternal night, taught by the heav'nly muse to venture down the dark decent, and up to reascend. . .
— John Milton

A few years ago, I had a heavy conversation with someone I came to know through various social circles. Over the years of casual hangouts, he opened up about his past. He was abused and neglected as a child and later spent five years in prison for assault with a deadly weapon and was charged as an adult. I felt a strange understanding for his aggression and built rapport through his vulnerability. What struck me most in that conversation was the recognition of callousness that manifests without restraint. He recounted the intensity of violence saying, "When you stab someone you don't quit when they tell you to stop; you quit when they stop telling you to." I could sense the hardening in his illustration, but what he further explained resonated on a deeper level:

"One time a fight broke out between two different gangs. I was on one side armed with a small shank. There was one guy from the other gang; he was the biggest guy they had and our biggest threat. He dropped forward after being kicked from behind and I managed to stab him before he could recover. I was scared as hell but kept fighting with pure adrenaline. I went to stab him again when he looked up toward me in pain. It was right then that I saw the same face that I gave to my mom growing up. It was like he spoke to me in a glance saying, 'I know how you feel, I was abused too.'"

At the core of our being is the need to be loved. A calloused heart is still a valuable heart to God. The enemy on the battlefield is hardly as frightening as the one inside ourselves, because often times that enemy is the externalized version of our own darkness. If we do not manage the battles within ourselves, they become the war the rest of the world must fight.

The aftermath of my own capacity for darkness became apparent when I returned home after the military. I remained cynical and bitter in civilized society as I had been under abnormal conditions. The environmental shift revealed the damage of my calloused state. The utility of diminished sensibilities was useful on deployment but isolated me from relationships at home. I was abrasive and short tempered. I resented people's innocence even though I was screaming for help inside. I was in danger of becoming what I fought.

People have told me I'm an approachable person. Those who know me well can point out my not so approachable side. Most people I meet today have no idea I even served, much less in the infantry. Even during my time in service, I received confused looks when I explained what I did for a living. My haircut was the only indicator. I didn't feel the need to give off the impression of a Marine. I learned to carry it inside, not as an identity. Most people in the United States will never experience what Marines are capable of overseas. Iraq thrusted us into the heart of darkness.

We swore in anger and spoke harshly in conversation. Our aggression was satiated in the company of peers. I developed a sense of courage and fearlessness around like-minded individuals. We were given the authority to stop people on the street, search vehicles and homes, and detain citizens against their will. There was nothing they could do about it. The authoritarian structure of the military entrusted us with the welfare of a nation. In a strange way, the freedom that came after my end of service felt more restrictive than the institution from which I departed. Without the company of Marines to channel my aggression I spiraled into repressed emotion.

The Marine Corps gave me a healthier image of myself, provided lasting relationships, and gave me a sense of purpose. But removed from a military context I was misaligned with all three pillars. I felt that I had learned things most people would never understand in their lifetime. Then, I was forced to abandon everything I had been taught in order to fit in. I became overly prideful in my accomplishments as a Marine. I felt that my status was above others and self-righteously felt I deserved respect for it. They didn't understand the purpose in "killing people" and often shied away without understanding what it had cost me. I was equipped to handle adversity in a violent context, but not the community of beneficiaries.

I routinely suppressed aggressive tendencies in jobs where my pillars suffered. My first job after the infantry was working overnight at a female clothing store restocking inventory for minimum wage. It was the only work I could find. I inquired about a manager's position, but was told I needed a Bachelor's degree. My experience leading junior Marines meant nothing. Likewise, in my next job, coworkers promoted above me on the basis of academic achievement. I felt degraded, being surrounded by other people who loitered into promotions through a traditional system.

I buried myself in academics thinking it would be the key to social and vocational success. Society held that a degree proved an employee's value. I registered for community college courses and began interacting with other students. The depth of our differences was vast. I was older and held contrasting views that offended people's sensibilities. I went on to receive my bachelor's degree with honors, but it didn't bring the social freedom I imagined it would. Academic endeavors offered some bonds of friendship but nothing as strong as those forged through adversity.

2 Corinthians 12:9-11 says, "But He said to me, 'My grace is sufficient for you, for my power is made perfect in weakness.' Therefore I will boast all the more gladly of my weaknesses, so that the power of Christ may rest upon me.'" I struggled with this verse in reference to my transition out of the military. In my pride it always seemed to mean I must handicap myself in order for God to be in charge. God does not negotiate with us. He seeks out those who willingly submit to his authority and uses the humble to gather the lost sheep.

Willingness is always more bearable than obligation. It is far better to choose humility than to be humbled by God. If God's power is perfected in our weakness, then our relationship with him is contingent upon submission. Submission is not about making myself less but learning to maintain obedience even if the world appears to be falling apart. Submission is about using what I've learned to elevate others in their weakness without capitalizing on it. Military service ends when a contract is completed. Servanthood is a lifelong commitment.

SERVING AFTER SACRIFICE

Do you see a man skillful in his work? He will stand before kings; he will not stand before obscure men.

— Proverbs 22:29

I was severely misaligned after the Marine Corps as most veterans find themselves to be. It took years to accept I may never form bonds like I did with the Marines of Weapons Company. It wasn't for lack of trying. Hardship and challenges forced us to rely on each other despite very real lifestyle differences. Many veterans have said, "I trust my brothers with my life but not my wife." Humans are tribal by nature. Without community we lose our sense of identity and purpose. My journey in civilian life led me to reassess my faith. I sought a place of belonging where I could begin a new journey.

After my first semester in college I was hired for a sales position with a company selling military and law enforcement equipment. The pay was decent and opened the door for stable employment. I gained full-benefits, paid time-off, a 401k plan, and was able to continue my education with flexible hours. I bought a house the following year and was able to sustain a mortgage while working and going to school. The structure of my post-military life was looking good. I felt I had found my place to thrive and connect with the people I worked with.

I worked my way into bigger projects and transitioned out of the sales department. Upward mobility within the company was promising until a lull in military contracts diminished the projects I was assigned to. Then, I was given menial tasks consisting of janitorial and construction work for the CEO. Myself and two other employees maintained the warehouse and the owner's residence as a way of staying busy and employable. I gave proposals for projects and offered solutions. None were accepted. I watched others promote over me though they were significantly less knowledgeable of the products we sold. After years with the company I became miserable and resentful.

Everyday became a struggle to get out of bed. I reflected on the Marine Corps and how I always felt connected to a purpose through the most miserable times. Then life became miserable and meaningless. Apathy pushed me to my breaking point. As I was driving home from work one day, I blasted hardcore music while screaming to the point of losing my voice. I reached over grabbing my company fleece and tore it to pieces in a rage. I had never felt so undervalued in my life. Aggression consumed me.

I collected myself at home and began developing a plan of action. Remaining with the company was no longer an option, but I had a mortgage to pay and a year left of school to complete my undergrad. Leaving the company would put me into a bad position financially and academically even though my mental health was declining. Feeling trapped, I sat before God in prayer. I refused to give into my crisis without knowing exactly how it would be resolved. I prayed for months and came to the realization that I needed to leave before securing another form of employment.

In my submission to God, I finally received an answer to my prayers. I placed my mental and spiritual health above my financial security. Leaving a toxic work environment for an unknown future felt insane. It was the next great unknown. I developed a budget to get through an entire year without work so I could finish my undergrad. I had no idea what I wanted to do for work but needed to find something by the end of the year. I managed to put together the funds necessary within four months and gave a two-week notice at the end of December that year. In addition to not knowing where God was leading me, I also felt him tell me not to seek other employment but to trust that he would provide.

A few months into the new year, a Marine Corps buddy told me about a shooting course taking place over the summer. It was a month long and eligible for the GI Bill. Seeing that I had the summer off without work or school, I signed up months in advance. Instead of sitting around for three months waiting out my last semester, I spent the last month of summer going through an extensive personal protection course learning how to guard high-profile individuals in high-risk environments. The course was geared for State Department contracting work but had other applications as well. I had no idea what I got myself into but enjoyed the camaraderie with other veterans. Though only a month long, I felt reconnected to a community of like-minded warriors.

The contrast between university students and veterans was eye opening in my final semester. Reuniting with a tribe for a short period of time reconnected me with a purpose and place to belong. I understood why I didn't fit in at a liberal arts university and wholeheartedly accepted it. Being around veterans reminded me to focus on my pillars even if that meant going seasons alone. I cherished my close friends in college with the knowledge that our lives would look completely different after graduation. College was a place that augmented my pillars and provided a temporary community. I still didn't know what I would end up doing at the end of the year but gained clarity on what it would consist of.

Two months before graduation my father forwarded me the contact information of a retired LAPD officer who worked doing executive protection. I hesitated to reach out in light of the promise I made to not seek employment during the year. With the end of the year approaching I decided to follow up with an opportunity that approached me. Because of my military background and recent training in personal protection, I was offered a position to work a high-profile client's estate doing security. I took the job and began working a month before my yearly budget would end. The company hired me at an hourly rate higher than my previous job, which took six years to reach.

Within six months, I received a pay and position increase as well as connected with another high-profile client to work their estates. After a year of praying through an unknown season I was now entrusted with the personal security of two Hollywood icons at double my previous yearly income. Ironically, the month-long course I took in between semesters did more for my future than a Bachelor's degree. Working in the protection industry has taken me across the United States and the world for various high-profile events, including working with the Secret Service next to presidential families, a Royal wedding, and numerous awards shows.

None of this was anything I asked for or imagined. I thrived in the executive protection industry surrounded by active and veteran law enforcement and military personnel who understood sacrifice. Additionally, I was never asked to submit academic credentials. I felt at home in my new community and duties. More importantly, it was an act of faith that made all of it possible. I was moved to tears when thinking back to where I was just a few years earlier. Frustration and apathy moved me from one industry into another. One season I'm hunting other men in a foreign country, then I'm folding panties and bras for minimum wage the next. In another season I'm mopping a warehouse floor, then walking a castle in Paris speaking broken French on behalf of an international icon. We rarely know where life will take us when we follow Christ.

In the final days of Jesus' life there is a profound moment when he exemplifies servanthood by washing his disciples' feet. John 13:6-9 captures Peter's interaction with Jesus. Peter refuses to be served in such a way, but Jesus replies, "'What I am doing you do not understand now, but afterward you will understand.' Peter said to him, 'You shall never wash my feet.' Jesus answered him, 'If I do not wash you, you have no share with me.'" Peter still has no idea what troubles will come in his life, but the promise remains despite his uncertainty.

I have been blessed in my journey. I don't believe Scripture is given to us as a promise of our prosperity in this life. But I do believe we will have a peace that surpasses all understanding if we are obedient to God's Word. Regardless of the status I've had the pleasure of working next to, there is only one true King. The God I trusted in the unknown of Ramadi is the God who entrusted me with the protection of modern royalty. He is the God who defines my pillars and is the author of my life. We are indebted to him because he paid the price of our rebellion on the cross. I attribute my worldly success to a persistent pursuit of God through faith in both feast and the famine. There is absolute power in prayer. There is absolute truth in God's Word.

MESSAGE IN METAPHOR

Fairy tales do not tell children the dragons exist. Children already know that dragons exist. Fairy tales tell children the dragons can be killed.

– G.K. Chesterton

I love great stories, particularly fiction. While I do read more non-fiction than fiction, something about fiction holds a different attention. The odd thing I began to notice in myself was the times I gravitated toward these types of stories in both books and movies. Since I got out of the military I've been reading tons of non-fiction. When I was deployed overseas in the heat of life's chaos and pain I was buried in fiction. This is something I didn't notice until recently. I never cared for books in school, but there is one that made me a reader to this day. Josh gave me a copy of the first Harry Potter novel on our first deployment in Iraq. I was reluctant to read it and made the excuse that I didn't want to read a kid's book. He persisted. I finished it within three days.

I read a good portion of it while on a patrol at the Government Center in Ramadi. At any moment I could have been hit by a mortar, RPG, or small arms fire, but I had grown beyond my fear of the environment back into the monotony of life. In order to avoid becoming a casualty of monotony, I ventured off into the world of fiction. When I started the book, it was the first time I read something purely for fun. It changed my perspective. I hadn't felt joy like that since before I joined the Marine Corps. My mind was transported from the 120-degree Ramadi desert to the foggy and mystical world of Hogwarts. For three days I traveled back and forth between reality and fiction as if the pages in front of me were a portal. I couldn't understand why I felt the way I did.

I returned the book to Josh after finishing it, and he looked at me with surprise. "Ooh, Mr. I don't want to read a kid's book!" He mocked me as I handed it back so soon. We talked about it and spent another day reflecting on a world outside of our own. Fiction reignited the creative part of my mind that had been murdered. I felt liberated. I sat in my own bewilderment at how something so simple could disconnect me from the apocalypse of my 19-year-old reality so easily. That set me on a path of a few other books during that deployment and the next. It didn't change reality, but it did provide a temporary distraction. I had never been fond of fiction before then, but in the most painful season of my life the made-up world made more sense than reality.

Movies were also a common form of relief in Ramadi. We all shared DVDs within the platoons and bought bootleg versions of current movies from Iraqis on base when we could. One that stuck out to me was Eternal Sunshine of the Spotless Mind. It was one of Jim Carrey's dramatic roles. The theme instantly captured my attention. The protagonist is fighting for a relationship that is inevitably doomed. I had to watch it a few times to get through the strange imagery, but there's a line toward the end that resonated deep.

In an ending scene, Joel's (Jim Carrey) relationship with Clementine (Kate Winslet) is coming to an end as he is having the memories of his love erased. The scene shows them sitting together on a beach waiting out their last memory together. As they wait out their last moment together Clementine asks what they should do. Joel's response was simply to enjoy it. Immediately I was reminded of the song that hit me after my first patrol. Movies, music, and books were therapeutic. They took my mind out of life's darkness because they revealed the deeper human story; not that it was possible to live through pain but that I could thrive in spite of it.

Stories are metaphors. They show us how ordinary people face extraordinary circumstances and come out the other side as better versions of themselves. The entertainment industry makes billions of dollars from it. Every fictional tale says something about us. They tell the human struggle for wholeness. The contexts, timelines, cultures, and languages all change, but the underlying portrayal of our own pain remains intact. Fiction allows us to detach from reality and step into a character in order to see hope rather than despair and that we might discover it in our own lives. Fiction is a metaphor for the human experience.

The paradox of a metaphor is that we often look for answers to specific problems when they only illustrate general principles. Metaphors entertain our creative brain but drop off when our rational brain needs to make decisions. We are left with the freedom to apply the principle or not in a given situation. This is our struggle with faith. Jesus often spoke in parables, which is a biblical term for a metaphor. He told parables in order to communicate to those seeking hope. He conveyed truth in a manner that would overcome the prideful intellect of rational minded individuals.

Metaphors are beautiful in books and movies but frustrating sometimes in Scripture. This is because entertainment focuses on us while Scripture is central to God's character. Some parables are easy to understand but others leave us searching for clarity. Jesus uses metaphors to illustrate God's will for humanity. His metaphors were misunderstood by those seeking their own advancement. It was those who were poor in spirit who were most receptive to his words. There is a historical usage of metaphors that goes back to humanity's oral history that pre-dates our written history. Humanity used metaphorical language long before anything was written down.

By the time humans started to document history, they wrote the stories that were told verbally for generations. Some argue that the book of Job (being the oldest book of the Bible) is not a literal story of a man but a metaphorical dialogue. The language of Job reflects a similar language style to dialogues in Greek Mythology which both originate in the 6th century B.C. The dialogue and narrative prose writing style tells us the story of Job in a way that conveyed a principle that would have been understood at that time. Regardless of the physical truths, the truths of Job's story reveal something about God's character. Scripture is comprised of various writing styles and metaphors. It is designed to reveal who God is so that we may understand who we are in light of his design and purpose.

Context plays another important role when applying the truths of Scripture. Jordan Peterson illustrates context with a unique analogy: Comparing an apple and an orange is difficult because they are not the same. So, what do you have? Two fruits. We generalize to a different level which places them in a category to be better understood together. What about an apple and a potato? You have two pieces of food. Here, a further generalization. What if you have an apple and a pencil? You have two objects. The highest category for things that have no connection. Generalizations allow us to contextualize at a level where things are loosely related—if at all—in order to make a connection.

Metaphors are generalizations that transcend the contextual differences of language, culture, industries, geography, and time. Parables communicate truths across time. Most who heard Jesus speak had a hard time understanding his parables. The disciples had a better understanding but still struggled. Luckily, they were close to him and he could explain the principles plainly. They didn't understand the reason for his language, but they eventually grasped onto his message. Before his death they believed he was the Savior that came to save them from the tyranny of Rome. After his death, they understood the eternal message he brought to the world. Many truths in Scripture are told through metaphors.

When Jesus spoke about the Prodigal Son and the Good Samaritan, he spoke to his community but communicated something to the world. It's humbling to read the words of Jesus 2,000 years later on the other side of the globe in a different culture and tongue and receive his words personally. Jesus spoke with an audience in mind, and that includes you and me. The truth is that we are in desperate need of Jesus. The fallen world we live in will always involve pain and suffering. The parables Jesus told were given in a fictional context where the embodied principles stand to guide us today even though we are far removed from their original audience. It is not difficult to rationalize the moral problem of evil on an intellectual level, but failure of faith is often found in the problem of pain. We are limited beings who respond primarily to our emotional states.

Historically, the Israelites experienced the signs and wonders of God and yet lived in rebellion against him. They did not struggle intellectually with God. They struggled in their emotional strain and obedience. In contrast, our modern American rebellion is driven by intellectual prestige and prosperity. We have descended into a culture of skepticism and doubt for God. Our emotional strain, unlike ancient Israel, is not driven by the refining holiness of God but the removal of his influence in our daily lives. Our culture is infected with an emotional arrogance that demands God explain himself in our suffering much like Job.

God responds in Job 38:2-3 saying, "Who is this that darkens counsel by words without knowledge? Dress for action like a man; I will question you, and you make it known to me." We will not always understand why we experience pain. Yet, through faith we are given hope that will not fail. Whether it is violence, the loss of a loved one, or the guilt we carry for our own actions, fulfillment is found when we seek to become like Christ. There is no greater love found in this life than in the person of Christ. God sent his only Son for reconciliation not for our temporal experience. It is through Christ that we move closer to the created order, as God intended the world to be. We

CALLOUSED HEART

are called to trust in him and not our own understanding. Until his return, we must learn to endure suffering, feed his sheep, and remain unstained from the world.

A FINAL PRAYER

God help us with the patience to seek You in all places. Give us the strength of a warrior not in fists or power, but in heart; to bear the weight of darkness that many fail to hold. Give us a spirit of perseverance to know the depth of love and all of its loss. We won't run away from You no matter how callous our hearts become. And we will remember to sit in silence to honor Your grace in that the ones we loved here, You first loved. No suffering is beyond You and You have the knowledge of all the world's grief. We can only hope that our reverence in prayer can hold even some of the burden You do. We are asking for more. We are asking to carry the weight of things most despised. Beyond anything You give us, we ask for the strength of Your Spirit to stand beyond trembling hands and now dry eyes. Our work is not done, and this moment with You is far from over.

Amen

AFTERWORD

There are endless opinions about the Global War on Terror (GWOT) and the United States' involvement in Iraq. I do not care to share my political convictions but do feel it is necessary to address the conclusion of OIF and OEF as faithfully as I can, while knowing politics are unavoidable. It was not my plan to write an afterword. However, I have been encouraged by many to do so in light of the events that concluded our longest war.

The 20th anniversary of the terrorist attacks on September 11th, 2001 arrived with an unprecedented burden of grief for Americans. The botched pull-out of US forces and the abrupt end to a twenty-year war was an act of political treason with seemingly no strategic oversight. The decision to end the longest war in American history was an honorable move while the execution unnecessarily cost lives and undermined thousands who sacrificed in pursuit of justice and peace. The withdraw from Iraq (OIF) is shamefully similar.

Ending the GWOT was long overdue, especially considering that both World Wars were fought in four years respectively. The prolonging of warfare leads to confusion, division, and poverty when nations do not have a hopeful future to move toward. I've seen veterans tearfully advocate for the welfare of Iraqi and Afghan nationals foreseeing the persecution that predictably follows from an abrupt end of conflict without full suppression of the enemy. One thing that should be clearly understood is that there is a distinct difference between service-members and politicians.

When I signed my contract after watching the Twin Towers fall, I was purely focused on combatting an enemy that brought the fight to our homeland. I was not given the choice to deploy to either Iraq or Afghanistan. I offered myself as a servant in service to his country, carrying an additional dog tag with the verse Isaiah 6:8 to remind me why I was fighting. Financial or economic incentives did not factor into my decision as there was no guarantee I would live to reap their benefits.

I'm often asked how I feel about the US' involvement in the Middle East having served there. Ethically speaking, the moral weight of this war rests on the politicians. The individual servicemembers who risked their lives executing orders have been faithful with what they were given. I do not wrestle with any attempts to justify my service or actions in the war. Neither should anyone else who served. When we all stand before God on judgement day we will all be accountable for what we have been entrusted with.

Anytime a moral dilemma is presented, I seek understanding in God's Word. Sometimes conflict is necessary. The intensity and context of any conflict are debatable and the only way to gauge a proper response is to begin and end with the very Word of God. If we have any hope of being a light in the world, we must be guided by Scripture. Conflict is unavoidable. Wars are sometimes unavoidable. The spiritual enemy is at war with God through creation. Until Christ comes to establish his kingdom, we will continue to live in a broken world.

It is conceivable that the next generation of service members will be involved in another conflict. It is certain that bad decisions will also be made. People will die and more injustice will permeate American history. We are delusional to think we are going to fix our broken world. The beauty of our American spirit is that we never give up. Regardless of how people distort our history, this nation was irrefutably founded on Christian principles. We set our eyes on the ideal and trip over ourselves as we attempt to live it out. This country will not last if we remove God from our daily lives. Ending the wars in Iraq and Afghanistan were the right decisions even though they were not executed with faith and discernment.

The conclusion of America's longest war is gravely disappointing but it is not debilitating. As a Christian, my focus is eternal. Everything we do holds weight in the next life. The author of Hebrews tells us to fix our eyes on Jesus who endured the cross for the joy to come (12:2). The Apostle Paul declares that our present sufferings do not compare to the glory to be revealed (Romans 8:18) and that love endures all things (1 Corinthians 13:7). All of the strife, all of the trauma, all of the political dissension, all of the social rejection should be seen as temporal attempts to take our eyes off Jesus. The Bible is abundantly clear of Christ's victory over the enemy and our adoption into the kingdom if we choose to believe in him. Hope belongs to those who fight the good fight and keep the faith.

ACKNOWLEDGEMENTS

To all those who contributed to this book both directly and indirectly. I cannot begin to express my gratitude for your time, expertise, and support throughout this process.

To my wife, Kendall – You have shown so much support for me through a myriad of emotional states as I wrestled with how to navigate this story, working long hours away from home, and making time for our relationship. I love you.

My parents – Your constant support and guidance has written this story. I cannot thank you enough as your generosity has given more than I could ever repay.

Charles "Sid" Heal – Thank you for your incredible insight on post-traumatic growth and the reality of combat from your own experiences since Vietnam. I cherish your kind words and humble spirit in taking time to share your wisdom with myself and future generations of service members and first responders.

Ken Rick – Leader of Marines. You exemplify the fighting spirit of the Corps. Your story and sacrifice are a model for the next generation of warriors who will carry on the legacy.

Linda Swanberg – Thank you for your undying support for the Marines of Weapons Company and your willingness to contribute to this work. I hope that our conversations behind the scenes and over the years have honoured Shane's memory and strengthened your faith in God.

ACKNOWLEDGMENTS

Julie Oligschlaeger – You have kept Chad's memory in our hearts for many years since we served. Thank you for your continued support of veterans and raising awareness about PTSD so others can find hope in their recovery.

Jin Park – For your spiritual guidance and not being afraid to challenge progressive ideas that have kept me stuck for years. You are a voice of reason, tempered with compassion, an amazing artist, Marine, and brother in Christ.

Charles Law – Your professional leadership and guidance over the years has been a blessing. The autonomy you've given me to work through the writing process has made this a reality.

Leah Brown – All the time you sacrificed to help me navigate a new arena of life. Without your guidance on marketing, social media strategy, publishing, and resources, this would not have been possible. You are an unexpected blessing and a beautiful soul.

My Editors – You have played a tremendous role in the development of this book, though you were too humble to be named or receive credit.

Thank you to all my co-workers who picked up work to make endless hours of writing possible. Many of you have proofread some of this work and offered thought provoking feedback and testimonials.

To all the Marines and service members who contributed to this work, I cannot thank you enough. This story is your story. You have all shared the burden of conflict and violence. I hope these words have been able to illustrate some of the emotions felt so others can begin to understand the sacrifice you all have made.

BIBLIOGRAPHY

Selected Books

Aquinas, Saint Thomas, and Dominicans. English Province. 1913. The *"Summa Theologica."* London, Burns Oates & Washbourne Ltd. [-42.

Baron-Cohen, Simon. 2022. *SCIENCE of EVIL: On Empathy and the Origins of Cruelty.* S.L.: Basic Books.

Beschloss, Michael. 2019. *Presidents of War.* S.L.: Broadway Books. p.369.

Brenneman, Todd M. 2014. *Homespun Gospel: The Triumph of Sentimentality in Contemporary American Evangelicalism.* Oxford; New York: Oxford University Press.

Browning, Christopher R, and Mazal Holocaust Collection. 2017. *Ordinary Men: Reserve Police Battalion 101 and the Final Solution in Poland.* New York: Harper Perennial.

Carson, D A. 2012. *Christ and Culture Revisited.* Grand Rapids, Mich.: William B. Eerdmans Pub. Co.

Cheney, Margaret. 1976. *The Coed Killer.* New York: Walker.

Chesterton, G K. (1908) 2020. *Orthodoxy.* S.L.: Blurb.

Conrad, Joseph, and Paul B Armstrong. 2017. *Heart of Darkness: Authoritative Text, Backgrounds and Contexts, Criticism.* New York: W.W. Norton & Company.

Dostoyevsky, Fyodor. 1950. *Crime and Punishment.* New York, Modern Library.

Dostoyevsky, Fyodor. 2019. *Brothers Karamazov.* S.L.: Echo Library.

Epictetus, and Thomas Wentworth Higginson. 1899. The Works

of Epictetus: Consisting of His Discourses, in Four Works, the Enchiridion, and Fragments. Boston: Little, Brown and Co.

Faber, David. 2010. *Munich, 1938: Appeasement and World War II.* New York: Simon Et Schuster.

Frankl, Viktor E, Ilse Lasch, Harold S Kushner, and William J Winslade. 2006. *Man's Search for Meaning.* Boston: Beacon Press.

Goethe, Von, and Carl Richard Mueller. 2004. Faust. Hanover, N.H.: Smith and Kraus.

Grossman, Dave, and Loren W Christensen. 2008. *On Combat: The Psychology and Physiology of Deadly Conflict in War and in Peace.* Millstadt, Il: Warrior Science Pub.

Grossman, David. 1996. *On Killing: The Psychological Cost of Learning to Kill in War and Society.* Boston: Little, Brown.

Henri J M Nouwen. 1979. *The Wounded Healer.* New York Doubleday (An Image Book).

Herndon, Booton, 1967. *The Unlikeliest Hero: The Story of Desmond T. Doss, Conscientious Objector, who Won His Nation's Highest Military Honor.* Boise, Idaho: Pacific Press Publishing Association. ISBN 978-0-8163-2048-6.

Junger, Sebastian. 2017. *Tribe.* Fourth Estate Ltd.

Leo Tolstoy, Graf. 2010. The Law of Love and the Law of Violence. Mineola, N.Y.: Dover Publications, Inc.

Loder, James E. 1998. *The Logic of the Spirit: Human Development in Theological Perspective.* San Francisco: Jossey-Bass Publishers.

Marcus, Aurelius, and Maxwell Staniforth. 1964. *Meditations. Translated with an Introduction by Maxwell Staniforth.* Harmondsworth, Eng: Penguin Books.

McCandless, Carine (2014). *The Wild Truth.* New York City: Harper One. ISBN 978-0-06-232514-3.

Milton, John. (1667) 2007. Paradise Lost: John Milton. Lewes

Gmc Distribution New York, N.Y. Spark Publishing.

Nietzsche, Friedrich Wilhelm, and Adrian Del Caro. 2014. *Beyond Good and Evil; on the Genealogy of Morality*. Stanford, California: Stanford University Press.

Owen, Roderic. 1978. *The Fate of Franklin*. London: Hutchinson.

Pasachoff, Naomi E.; Littman, Robert J. (2005) [1995]. *A Concise History of the Jewish People*. Reference, Information and Interdisciplinary Subjects Series. Lanham, Maryland: Rowman & Littlefield. p. 67. ISBN 9780742543669. Retrieved 4 February 2021

Peterson, Jordan B. 2017. *Summary of 12 Rules for Life: An Antidote for Chaos*. Richdad Summaries.

Rudyard Kipling, and Thomas Stearns Eliot. 1990. *A Choice of Kipling's Verse*. London: Faber And Faber.

Shakespeare, William, William George Clark, and William Aldis Wright. 2009. Macbeth. Charleston, Sc: Bibliolife.

Søren Kierkegaard, Victor Eremita, and Alastair Hannay. 2004. *Either/Or: A Fragment of Life*. London Penguin Books.

Spurgeon, Charles H., Gilbert, Josiah H. 1895. *Dictionary of Burning Words of Brilliant Writers: A Cyclopædia of Quotations from the Literature of All Ages*. New York: W.B. Ketcham.

Stone, Michael H. 2009. *The Anatomy of Evil*. Amherst, New York: Prometheus Books.

Thoreau, Henry David. (1854) 2016. Walden. Canterbury Classics. Original Publication, Ticknor and Fields: Boston.

Vronsky, Peter. 2004. *Serial Killers: The Method and Madness of Monsters*. New York: Berkley Books.

Wheeler, John. 1954. *The Nemesis of Power, the German Army in Politics, 1918-1945. John W. Wheeler-Bennett*. London: Macmillan.

Websites

Brennan, Margaret. 2019. "Full Transcript of 'Face the Nation' on September 8, 2019." Www.cbsnews.com. CBS Interactive Inc. September 8, 2019. https://www.cbsnews.com/news/full-transcript-of-face-the-nation-on-september-8-2019/.

Brom, Robert. 2018. "The Galileo Controversy." Catholic Answers. 2018. https://www.catholic.com/tract/the-galileo-controversy.

Cocca, Christina. 2014. "Isla Vista Rampage Almost 'Impossible to Prevent': Napolitano." NBC Los Angeles. NBC Universal Media. May 26, 2014. https://www.nbclosangeles.com/news/isla-vista-rampage-uc-president-janet-napolitano/70084/.

Cowell, Alan. 1992. "After 350 Years, Vatican Says Galileo Was Right: It Moves." *The New York Times*, October 31, 1992. https://www.nytimes.com/1992/10/31/world/after-350-years-vatican-says-galileo-was-right-it-moves.html.

Crecente, Brian. 2008. "Gamer Was on Deadly Road: Local News: The Rocky Mountain News." Web.archive.org. Scripps Newspaper Group Online. June 22, 2008. https://web.archive.org/web/20080622212021/http://www.rockymountainnews.com/drmn/local/article/0%2C1299%2CDRMN_15_4722344%2C00.html.

Felton, David. 1970. "Charles Manson: The Incredible Story of the Most Dangerous Man Alive." Rolling Stone. Rolling Stone. June 25, 1970. https://www.rollingstone.com/culture/culture-news/charles-manson-the-incredible-story-of-the-most-dangerous-man-alive-85235/.

Follman, Mark, and Becca Andrews. 2015. "Here's the Terrifying New Data on How Columbine Spawned Dozens of Copycats." Mother Jones. October 5, 2015. https://www.motherjones.com/politics/2015/10/columbine-effect-mass-shootings-copycat-data/.

Glenn, Cameron, Mattisan Rowan, John Caves, and Garrett Nada. 2016. "Timeline: The Rise, Spread and Fall of the Islamic State." Wilson Center. 2016. https://www.wilson center.org/article/timeline-the-rise-spread-and-fall-the-islamic-state.

Kington, Tom. 2013. "Vatican Offers 'Time off Purgatory' to Followers of Pope Francis Tweets." The Guardian. July 16, 2013. https://www.theguardian.com/world/2013/jul/16/ vatican-indulgences-pope-francis-tweets.

Otten, Joseph. 1911. "CATHOLIC ENCYCLOPEDIA: Musical Instruments in Church Services." Www.newadvent.org. October 1, 1911. https://www.newadvent.org/cathen /10657a.htm.

Peterson, Jordan B. 2020. "The Jordan B. Peterson Podcast: Jean Piaget (Constructivism)." Podcast, S3E30. Apple Podcasts App. https://www.jordanbpeterson.com/podcast/.

R.C. Sproul. 2014. "Questions and Answers #2 by Various Teachers." Ligonier Ministries. Ligonier Ministries. 2014. https://www.ligonier.org/learn/conferences/overcoming-the-world-2014-national-conference/questions-and-answers-2-2014-national/.

Rick, Kenneth. 2019. "Out from under the Rug." *Leatherneck Magazine*, August. https://mca-marines.org/wp-content /uploads/Outfromundertherug.pdf.

Slick, Matt. 2015. "What Are the Main Divisions of the Old Testament Law?" Christian Apologetics & Research Ministry. February 19, 2015. https://carm.org/about-doctrine/what-are-the-main-divisions-of-the-old-testament-law/.

Smith, Jay. 2019. "Job Summary." Biblehub.com. Bible Hub. 2019. https://biblehub.com/summary/job/1.htm.

Spafford, Horatio, Anna Spafford, and Currier & Ives. 2005. "Family Tragedy - the American Colony in Jerusalem | Exhibitions - Library of Congress." Www.loc.gov. Library of Congress. January 12, 2005. https://www.loc.gov/exhibits/americancolony/amcolony-family.html.

Stewart, Phil. 2016. "Trump to Nominate Retired General Mattis for Pentagon." *Reuters*, December 2, 2016. https://www.reuters.com/article/us-usatrumpdefenseid USKBN13Q5RU.

Thomas, Kristen. 2018. "The Murder of 'Kitty' Genovese That Led to the Bystander Effect & the 911 System." *The Vintage News*, June 8, 2018, sec. Archives. https://www.thevintagenews.com/2018/06/08/kitty-genovese/.

Wes. 2015. "Tubal Cain." Masonic Beehive. May 8, 2015. https://masonicalvarium.com/2015/05/08/tubal-cain/.

Wikipedia Contributors. 2019. "Desensitization (Psychology)." Wikipedia. Wikimedia Foundation. October 2, 2019. https://en.wikipedia.org/wiki/Desensitization_(psychology).

Yale Law School. 2019. "The Avalon Project: Munich Pact 9/29/38." Yale.edu. Lillian Goldman Law Library. 2019. https://avalon.law.yale.edu/imt/munich1.asp.

Video

Baucham, Voddie. "Voddie Baucham on Homosexuality." *Www.youtube.com*, Peter Salas, 28 Jan. 2019, www.youtube.com/watch?v=HfbxxFA_oP4. Accessed 9 July 2021.

Bourgoin, Stéphane. "Ed Kemper Interview - 1991 (Extended)." *YouTube*, 8 July 2016, www.youtube.com/watch?v=j8IfslxOmF0. Accessed 2 June 2021.

Boyd, Greg. "The Non-Violent Warrior." *Www.youtube.com*, Woodland Hills Church, 3 Apr. 2017, www.youtube.com/

watch?v=VXF61JkfPlQ. Accessed 22 Apr. 2021.

Eternal Sunshine of the Spotless Mind. Directed by Michel Gondry, Focus Features, 19 Mar. 2004.

The Meeting House. "Cruciform Theology with Greg Boyd & Bruxy Cavey L the Meeting House after Party." *YouTube*, 18 June 2019, www.youtube.com/watch?v=cBj0hebeWCY. Accessed 22 Apr. 2021.

Various

Bible Hub. "2 Timothy 3:3 Amplified Version." *Biblehub.com*, Bible Hub, 2019, biblehub.com/2_timothy/3-3.htm. Accessed 21 Jan. 2021.

Merriam-Webster.com Dictionary, s.v. "callous," accessed July 9, 2021, https://www.merriam-webster.com/dictionary/callous.

Merriam-Webster.com Dictionary, s.v. "pacifism," accessed April 24, 2018, https://www.merriam-webster.com/dictionary/pacifism.

Rick, Kenneth. *Sergeant Rick Awarded the Silver Star.* 23 May 2021.

Sanderson, Steve. "Iraq Journal #1." 2005.---. "Iraq Journal #2." 2007.

Smith, Kevin. "Behavioral Threat Assessment: Preventing the Active Shooter." Received by Steve Sanderson, 3 Mar. 2020.

Uszak, Erica L., ""Peace for Our Time": Past and Present Receptions of Neville Chamberlain's Speech and the Munich Agreement" (2020). Student Publications. 793. Retrieved 9 July, 2021.

PERMISSIONS

ABOUT THE AUTHOR

STEVE J SANDERSON joined the Marine Corps in 2004 and served with 3rd Battalion 7th Marines, Weapons Company until 2008. He was deployed three times with two combat deployments to Ramadi, Iraq and a combined-arms training deployment in South Korea. After the Marines Steve pursued a degree in Psychology and worked in corporate sales for military and law enforcement agencies. He is currently an Executive Protection agent for high-profile clients in southern California. Steve is also a musician and has served on various worship teams since high school. His faith in God has been the most influential part of his life and has helped him in his post-traumatic growth.

Did you enjoy this book?
I'd love to hear what you thought about it!
authorstevejsanderson@gmail.com

Let's connect:

@STEVEJSANDERSON